Lineworker Rigging Practices

LINE CONSTRUCTION
ADVANCEMENT FUND
Powered by Northeast Line Contractors

in partnership with

American Technical Publishers, Editorial Staff

Editor in Chief:
    Jonathan F. Gosse
Vice President—Production:
    Peter A. Zurlis
Assistant Production Manager:
    Nicole D. Bigos
Digital Media Coordinator:
    Adam T. Schuldt
Art Supervisor:
    Sarah E. Kaducak
Technical Editor:
    Scott C. Bloom
Copy Editor:
    Talia J. Lambarki

Editorial Assistant:
    Alex C. Tulik
Cover Design:
    Nicole S. Polak
Illustration/Layout:
    Nicole S. Polak
    Robert M. McCarthy
    Thomas E. Zabinski
    Nick W. Basham
    Bethany J. Fisher
    Sarah E. Kaducak
Digital Resources
    Cory S. Butler
    Heather M. Parker

1 2 3 4 5 6 7 8 9 – 15 – 9 8 7 6 5

Printed in the United States of America

ISBN 978-1-935941-21-7

 This book is printed on recycled paper.

Acknowledgements

This book has been developed with support provided by the Line Construction Advancement Fund.

Powered by Northeast Line Contractors

The authors and publisher are also grateful to the following companies and organizations for providing photographs and information:

AirFloat
Altec, Inc.
Bishop Lifting Products
Brooks Brothers Trailers
Buckingham Manufacturing Company, Inc.
The Crosby Group, LLC.
Erickson Incorporated
Harrington Hoists, Inc.

Hilman Incorporated
Klein Tools, Inc.
Lift-All Company, Inc.
Skarnes, Inc.
Southern California Edison
Southwire Company, LLC.
Tindall Corporation

Textbook Development Committee:

Dan Dade
Mike Fitzpatrick
Jason Iannelli
Don Jamison

James McGowan
Virgil Melton
Bill Stone

Contents

Contents

LEARNER RESOURCES

Quick Quizzes™
Illustrated Glossary
Flash Cards

Crane Hand Signaling Quiz
Media Library
ATPeResources.com

Introduction

Lineworker Rigging Practices is an introduction to the physical principles, safety considerations, and common practices involved in hoisting loads in the outside line industry. This involves planning a lift, evaluating and preparing a load, choosing appropriate rigging equipment, rigging a load, communicating hoisting instructions, and performing a lift safely. Each step is critical to maintaining a controlled lift with a proper margin of safety. This book also includes instruction on the following critical skills and procedures:

- inspecting and maintaining rigging equipment
- comparing equipment ratings for different sling types, sizes, and hitches
- calculating sling loads
- tying practical knots, hitches, and splices
- calculating load weight
- determining a load's center of gravity and balance points
- evaluating the mechanical advantage of block-and-tackle assemblies
- signaling a digger derrick or crane operator for desired crane motions

When used in conjunction with the appropriate hands-on training, *Lineworker Rigging Practices* provides a solid foundation of safe material handling practices for lineworkers. Further study may lead to specialization in rigging and crane operation.

Objectives identify the main concepts in the chapter to be mastered.

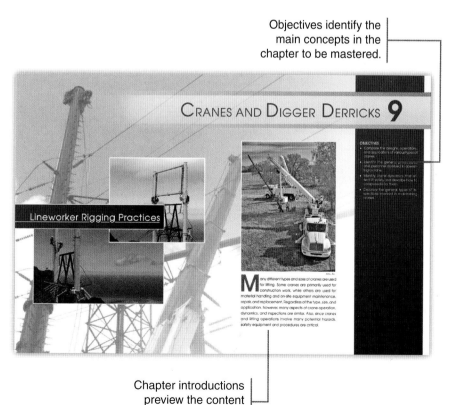

Chapter introductions preview the content to be covered.

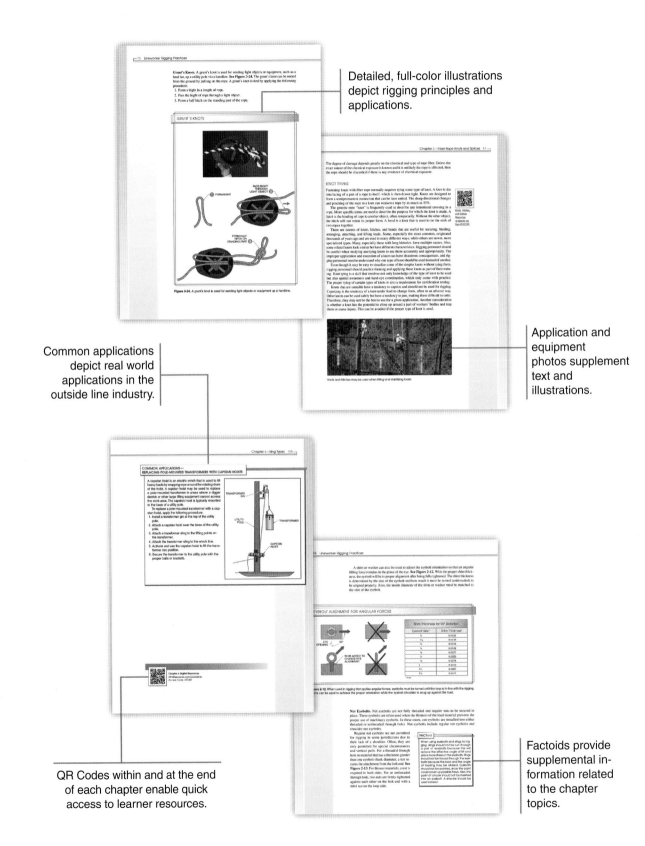

Detailed, full-color illustrations depict rigging principles and applications.

Application and equipment photos supplement text and illustrations.

Common applications depict real world applications in the outside line industry.

QR Codes within and at the end of each chapter enable quick access to learner resources.

Factoids provide supplemental information related to the chapter topics.

Learner Resources

Lineworker Rigging Practices includes a variety of online learner resources that enhance chapter concepts and promote learning.

Quick response (QR) Codes, located at the end of each chapter and in selected locations throughout the book, provide easy access to online learner resources that enhance chapter concepts and promote learning. These resources can be accessed by either of the following methods:

- Go to qr.njatcdb.org and enter the appropriate item number.
- Use a QR Code reader app to scan the QR Code with a mobile device.

For additional information, visit qr.njatcdb.org Item # 1649

The Learner Resources include the following:

- **Quick Quizzes**™ that provide interactive questions for each chapter, with embedded links to highlighted content within the textbook and Illustrated Glossary
- An **Illustrated Glossary** that defines terms and includes links to selected textbook illustrations
- **Flash Cards** that provide a self-study/review tool for identifying rigging terms and definitions, common rigging hardware, types of crane motions, standard hand signals for directing crane motions, and practical knots, hitches, and splices
- A **Crane Hand Signaling Quiz** that tests knowledge of hand signals required for specific crane motions
- A **Media Library** that contains videos and animations that reinforce and expand textbook content
- **ATPeResources.com,** which provides access to online reference materials for continued learning

Lineworker Rigging Practices

INTRODUCTION TO RIGGING

OBJECTIVES

- Explain rigging and hoisting as used in the outside line industry.
- Describe the unique aspects of personnel hoisting.
- Describe the major safety considerations of hoisting activities.
- List the major regulations and standards related to rigging and hoisting activities.
- Explain the importance of understanding OSHA personnel definitions and rating terminology.
- Explain how to acquire certification through training and examination.

Rigging and hoisting are common and often critical parts of outside line work. They are also potentially hazardous activities, so all personnel involved must be well trained and experienced. These personnel include outside linemen, crane operators, riggers, signalpersons, and any auxiliary personnel needed to assist with the operation and ensure the safety of all involved. Knowledge of the relevant regulations and standards for rigging and hoisting is required for all involved personnel. Certification in these qualifications may also be required.

RIGGING AND HOISTING IN THE OUTSIDE LINE INDUSTRY

Hoisting is the transportation of a load by a crane or hoist. This is also commonly called lifting. A *load* is an object that must be transported via hoisting. A crane or hoist is attached to a load by means of rigging. *Rigging* is the assembly of components that connect a load to the lifting hook of a crane or hoist. **See Figure 1-1.** A *lift* is one complete set of hoisting actions performed by workers who transport a load from a starting point to a destination.

Rigging and hoisting operations are used to complete a variety of lifts in the outside line industry. Some lifts are small, simple lifts of a few hundred pounds, while others are heavy and complex lifts involving thousands of pounds. For example, the largest types of cranes are typically used for erecting transmission towers.

The personnel involved in rigging and hoisting require extensive training in safety and operations. Because each lift may be different, personnel must be able to easily adapt to different loads, crane types, and hoisting procedures. Also, given the relatively large scale of most lifts, multiple people are usually involved, each directing a certain aspect of the operation. Separate training and/or certification may be required for each job.

RIGGING AND HOISTING

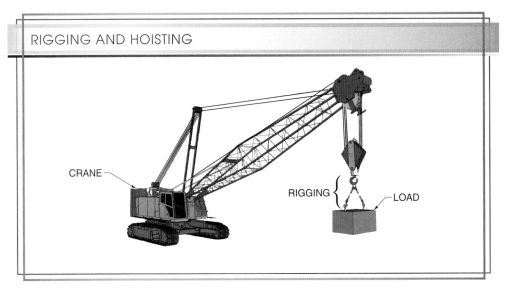

CRANE

RIGGING

LOAD

Figure 1-1. Hoisting is a method of transporting a load with a crane or hoist. Rigging is used to attach a load to a crane.

Outside Line Construction

Outside line construction includes many rigging and hoisting operations, such as hoisting tools, equipment, transformers, conductors, utility poles, and transmission towers. Digger derricks, material-handling aerial devices, cranes, capstan hoists, and block and tackle systems can be used to perform these operations. Digger derricks are vehicles that are often used to hoist and set utility poles. **See Figure 1-2.** Material-handling aerial devices are used to hoist transformers and other equipment. Cranes are used to offload large construction materials from delivery vehicles, transport materials around a job site for fabrication or processing, and set finished components in place. A *capstan hoist* is a portable winchlike device with a rotating drum (capstan) used for lifting. *Block and tackle* is a combination of sheaves and ropes that improve lifting efficiency.

OUTSIDE LINE CONSTRUCTION HOISTING

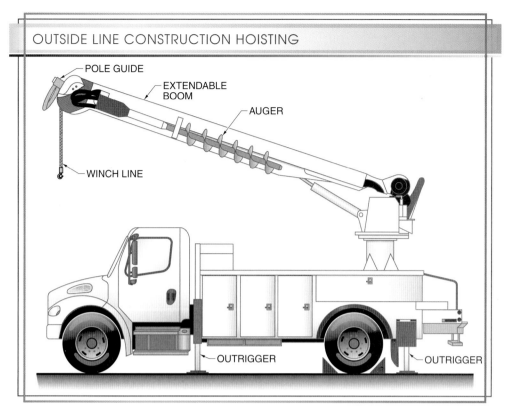

Figure 1-2. Digger derricks are used for various hoisting applications in outside line construction.

Maintenance

Rigging and hoisting are also used for a variety of maintenance activities, such as moving line tools or equipment to the work area and replacing equipment. For example, a rigging and hoisting operation can be performed to remove a damaged transformer from a utility pole and replace it with a new transformer. **See Figure 1-3.**

MAINTENANCE HOISTING

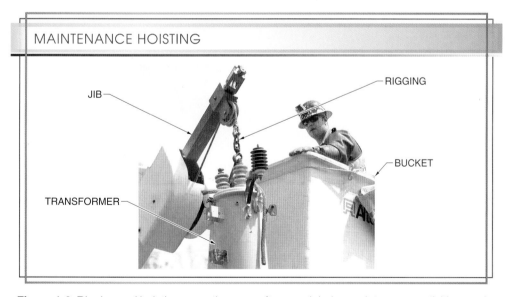

Figure 1-3. Rigging and hoisting operations are often used during maintenance activities, such as repairing and replacing equipment.

Because the needs for rigging and hoisting during maintenance activities can vary widely, personnel must be well trained to handle many different situations. For example, personnel must be able to select the correct rigging components, connect a load properly, and hoist a load safely. However, most loads are likely to be small enough for one or two people to perform the job. Large lifts may require greater preparation, and additional personnel may be necessary.

Personnel Hoisting

Cranes and material-handling aerial devices can be used to hoist outside linemen in a bucket or cage. Outside linemen must be positioned at an elevated location to perform work. Typically, outside linemen stay in the bucket or cage until their task is completed. **See Figure 1-4.**

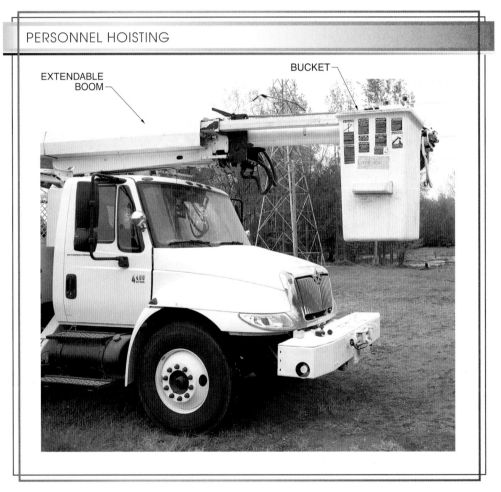

PERSONNEL HOISTING

EXTENDABLE BOOM

BUCKET

Figure 1-4. Material-handling aerial devices are used for personnel hoisting.

Hoisting outside line personnel is different from rigging and hoisting loads. An additional and extensive set of safety regulations are applied, and special training is required. While rigging and hoisting loads involve some risk to personnel on the ground, the risk to hoisted personnel is far greater. Any failure of the rigging or hoisting equipment brings the immediate risk of death or serious injury to those personnel. Also, the actions of the hoisted personnel can be an additional hazard to both themselves and those on the ground.

For example, workers may forget to attach their safety lines, or tools and equipment may be more likely to fall when used. For these reasons, many hoisting publications and training programs treat the hoisting of personnel as a separate and advanced topic. Proper instruction and testing must be completed by those involved in hoisting personnel, often after suitable training in basic rigging, hoisting, and signaling practices.

SAFETY

Rigging and hoisting activities have a long history, and the related safety practices have evolved significantly over time. Initially, experience along with trial and error led to rules of thumb, which were passed on via formal and informal peer-to-peer training. Over time, the safety for rigging and hoisting has advanced to scientific approaches. Modern safety practices have dramatically decreased the rates of accidents and fatalities, yet many still occur due to unsafe conditions and actions.

Hazards

While the most common hazard to personnel during hoisting is being struck by a falling object, hoisting-related accidents can cause many different types of injuries. Accidents may involve falls, contact with electric current, and being caught between parts of a crane or between a part of a crane and another object. Transportation and material-handling accidents involving hoists also contribute to worker injuries and deaths. Even using manual- or power-operated hoists that are not part of a crane carries a significant hazard potential.

The potential for damage to equipment, or injury or death to involved personnel or to bystanders, is a very real possibility during hoisting operations. Furthermore, it is possible to successfully complete a hoisting operation only to have a serious accident occur due to incidental slipping, falling, or improper material handling. A comprehensive approach to safety should be taken by all personnel in every aspect of their employment, especially when dealing with rigging and hoisting duties.

Personnel Training

Many safety practices are simply common sense. Keeping the workplace clean and looking out for easily recognizable hazards is a good starting point. However, the complex situations and operations present in outside line rigging and hoisting require special training in order to identify and avoid potential hazards. Safety awareness and training programs, along with job site evaluation and enforcement, can help establish a high level of workplace safety. Ongoing efforts in training, employing new techniques and materials, and assessing outcomes can help maintain safety. Even veteran workers may require additional training to deal with the safety issues presented by new tasks.

Newly trained personnel typically have a greater awareness for potential safety issues but may lack the field experience to develop effective solutions. Experienced personnel have a tendency to rely on established procedures, but these procedures may no longer be appropriate. Achieving a balance of up-to-date technical training and practical field experience can be a challenge for employers and employees alike. Combining ongoing technical study and hands-on field exercises helps provide personnel with the skills needed for specific job tasks.

Personnel engaged in rigging, hoisting, and signaling activities must be qualified, but federal regulations leave it to employers to determine a person's qualifications. Most employers rely on third-party organizations to administer standardized training and exams and certify that the student has mastered a minimum skill level.

Equipment Safety

Regular inspection of all rigging and hoisting equipment must be part of a safety program. There are general guidelines for performing frequent and periodic inspections for the most widely used pieces of equipment. However, it should be kept in mind that equipment manufacturers may have inspection requirements that are specific to particular types of equipment. It may be necessary to consult a manufacturer's service manuals to inspect and service some equipment.

> **FACT**oid
>
> Misuse of equipment is the leading cause of equipment failure, and can cause injuries. It is important to follow manufacturer recommendations, such as making level lifts, to avoid unsafe use of equipment. During a level lift a hoist hook must be located directly above the center of gravity. If the hook is too far on either side, dangerous tilting will result.

The installation and use of specialized rigging and hoisting equipment may also require workers to consult a manufacturer's installation and usage instructions. These instructions should be followed in lieu of the general safety procedures provided in safety standards, provided that they do not conflict with the standards.

Lift Planning

Lift planning is a critical part of a comprehensive rigging and hoisting safety program. Many poorly planned jobs have resulted in catastrophic failures, damage to equipment, and injury or death to personnel.

Every lift should be planned beforehand, regardless of how simple it seems to be. First, the overall hoisting operation must be considered, including where and how the load will be unloaded, stored, moved, and hoisted into place. Second, a list of required rigging and material handling equipment should be compiled. All equipment should be procured prior to material-handling and hoisting activities so there are no delays that might encourage makeshift solutions. Finally, the difficulty of the lift should be evaluated along with the skill levels of the personnel involved. Complex lifts may require additional training or more experienced personnel.

Fatality Assessment and Control Evaluation (FACE) Program

Regrettably, despite numerous safety regulations and programs, accidents still occur, and some are fatal. Every day in the United States, an average of 16 workers die as a result of workplace accidents. These incidences are investigated to determine the cause of each accident and to recommend strategies to avoid similar problems in the future. This information is used to continually update and improve safety regulations and standards. Some reports are also made available publicly so safety professionals, industry leaders, contractors, and workers can recognize similar weaknesses in their safety programs and quickly remedy them.

One source for this information is the Fatality Assessment and Control Evaluation (FACE) program, administered by the National Institute for Occupational Safety and Health (NIOSH). The FACE program targets accidents in construction and related industries. On-site investigations are conducted to collect facts and witness observations from accidents, which are compiled into comprehensive FACE reports. **See Figure 1-5.** Researchers use this information to identify new hazards, which may suggest the need for new research or prevention efforts or for new or revised regulations to protect workers. The individual reports and interpretive publications are disseminated to relevant audiences and are available online through the NIOSH home page.

For additional information, visit qr.njatcdb.org Item #1647

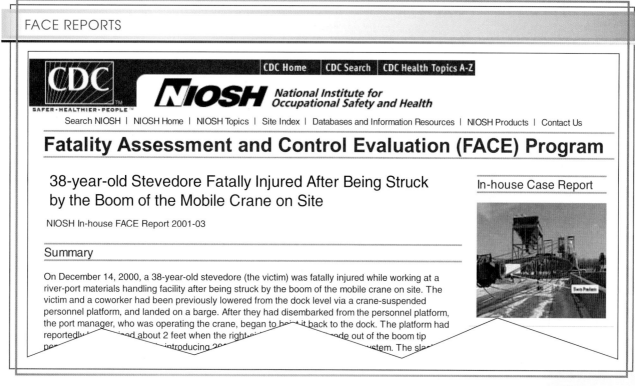

FACE REPORTS

CDC Home | CDC Search | CDC Health Topics A-Z

NIOSH *National Institute for Occupational Safety and Health*

Search NIOSH | NIOSH Home | NIOSH Topics | Site Index | Databases and Information Resources | NIOSH Products | Contact Us

Fatality Assessment and Control Evaluation (FACE) Program

38-year-old Stevedore Fatally Injured After Being Struck by the Boom of the Mobile Crane on Site

In-house Case Report

NIOSH In-house FACE Report 2001-03

Summary

On December 14, 2000, a 38-year-old stevedore (the victim) was fatally injured while working at a river-port materials handling facility after being struck by the boom of the mobile crane on site. The victim and a coworker had been previously lowered from the dock level via a crane-suspended personnel platform, and landed on a barge. After they had disembarked from the personnel platform, the port manager, who was operating the crane, began to hoist it back to the dock. The platform had reportedly been raised about 2 feet when the right-side [...] de out of the boom tip [...] introducing 2[...] stem. The sla[...]

Figure 1-5. Fatality Assessment and Control Evaluation (FACE) reports are widely available so others can develop regulations and change procedures in order to prevent similar accidents from happening.

REGULATIONS AND STANDARDS

Many government agencies and independent organizations help ensure safe work environments by developing rules regarding safety equipment and procedures. Some rules are mandatory, and some are voluntary. A *regulation* is a rule made mandatory by a federal, state, or local government. A *code* is a collection of regulations related to a particular trade or environment.

For additional information, visit qr.njatcdb.org
Item #1648

Occupational Safety and Health Administration (OSHA) Regulations

In the United States, the Occupational Safety and Health Administration (OSHA) is the federal agency charged with ensuring workplace safety. OSHA develops and publishes Title 29 of the Code of Federal Regulations (CFR), which includes the regulations that cover workplace safety.

The most widely referenced workplace safety regulations in OSHA 29 CFR are Parts 1910 and 1926, which apply to general industry and construction, respectively. Additional OSHA regulations apply specifically to line work, such as 29 CFR 1910.269 – *Electric Power Generation, Transmission, and Distribution* and 29 CFR 1926 Subpart V – *Power Transmission and Distribution*. Also, the construction regulations in Part 1926 often reference information from the general industry regulations in Part 1910 that deal with specific safety requirements.

Rigging and hoisting personnel must be familiar with all relevant OSHA regulations, which include not just rigging and hoisting but also material handling and storage, tools, signs, signals, and barricades, at a minimum. **See Appendix.** Fall protection and other personal protective equipment (PPE) are also areas of OSHA regulation that may be required in the course of performing rigging and hoisting operations.

OSHA regulations are the minimum requirements for safety in the workplace, and employers are free to require employees to conform to more rigorous standards. Many employers and industry groups have developed their own safety standards and procedures to address specific problems within their industries and may require specialized training for those performing certain duties. OSHA regulations describe the requirements for workplace safety without prescribing how any given task should be accomplished. Frequently, there are a number of ways to perform tasks safely, and additional, relevant safety standards can provide direction on specific safety practices.

Standards

A *standard* is a collection of voluntary rules developed through consensus and related to a particular trade, industry, or environment. Standards are developed and published by the members of standards organizations, often industry-specific, which periodically review and refine the rules in the standards. Standards themselves have no authority unless they are adopted by a federal, state, or local government as regulations. Then they become mandatory and enforceable by adopting that government unit.

ASME B30 Standard. The American Society of Mechanical Engineers (ASME) is one of the primary coordinating entities for standards development for rigging and hoisting. ASME standards deal with specific requirements for the selection, inspection, and utilization of cranes and material-handling equipment. These requirements are referred to as the B30 Standard, which contains 29 subparts. **See Appendix.** The requirements within the B30 Standard are updated periodically in order to incorporate new knowledge, equipment, and techniques. Generally, the most recent version of any standard should be used.

Some subparts in the B30 Standard are only relevant to specific situations, but others are the basis for most rigging safety and training requirements. The most commonly referenced subsections of the standard for everyday rigging tasks include the following:
- B30.5 *Mobile and Locomotive Cranes*
- B30.9 *Slings*
- B30.10 *Hooks*
- B30.12 *Handling Loads Suspended from Rotorcraft*
- B30.16 *Overhead Hoists (Underhung)*
- B30.20 *Below-the-Hook Lifting Devices*
- B30.21 *Lever Hoists*
- B30.26 *Rigging Hardware*

ASSE A10.42 Standard. Another standard relevant to rigging and hoisting, published by the American Society of Safety Engineers (ASSE), is ANSI/ASSE A10.42, *Safety Requirements for Rigging Qualification and Responsibilities*. This standard, which pertains to construction and demolition operations, deals with personnel qualification and training as well as the performance requirements for riggers under normal and critical operations.

DOE STD-1090 Standard. The U.S. Department of Energy (DOE) publishes DOE STD-1090, *Hoisting and Rigging*. This document provides guidelines for hoisting and rigging operations within power-generating stations in the United States. A unique feature of the DOE standard is that it requires all lifts to be categorized as ordinary, critical, preengineered production, or personnel before the lift is planned and executed. The systematic approach used in this standard can be a valuable tool not only in the power-generating industry but also for lifting operations in other industries.

National Electrical Safety Code®. The National Electrical Safety Code® (NESC®) is a standard published by the Institute of Electrical and Electronics Engineers (IEEE) that is used throughout the outside line industry. The NESC provides a set of rules for safeguarding persons during the installation, operation, and maintenance of electric supply and communication lines and associated equipment.

Terminology

Terminology and its specific meaning becomes critical when dealing with both enforceable regulations and a potentially hazardous field operation, such as hoisting. Without a clear understanding of certain words, even a simple task can be dangerous.

OSHA Personnel Definitions. OSHA regulations frequently require that certain duties be performed by either a competent person, a qualified person, or in some cases by a person who is deemed to be both competent and qualified. A *competent person*, according to OSHA, is "one who is capable of identifying existing and predictable hazards in the surroundings or working conditions which are unsanitary, hazardous, or dangerous to employees, and who has authorization to take prompt corrective measures to eliminate them." A *qualified person*, according to OSHA, is "one who, by possession of a recognized degree, certificate, or professional standing, or who by extensive knowledge, training and experience, has successfully demonstrated the ability to solve or resolve problems relating to the subject matter, the work, or the project." When reviewing OSHA requirements, an individual must keep these definitions in mind and ensure that the proper personnel are involved in a task.

Rating Terminology. Despite the extensive body of regulations and standards covering rigging, hoisting, and signaling practices, there is still inconsistency throughout the industry for the terminology defining the amount of force that a component may be subjected to and still maintain the appropriate margin of safety. No single term stands out as an industry standard, though "rated load" is used by many ASME standards for rigging components. Also, there is a slight preference among crane and hoist manufacturers for "rated capacity." Other common terms include "load rating," "load capacity," "weight rating," "weight capacity," "working load limit (WLL)," and "safe working load." **See Figure 1-6.** These terms all have the same definition and, when encountered in the field or in publications, can be treated as equivalent.

RATING TERMINOLOGY

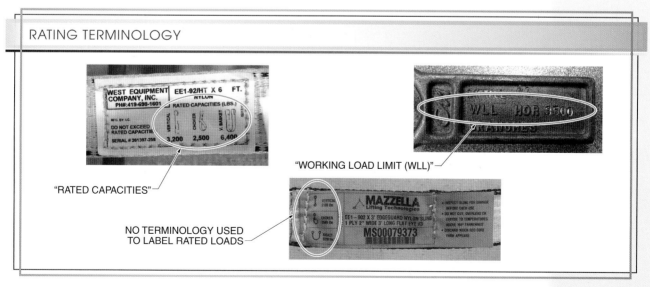

"RATED CAPACITIES"

"WORKING LOAD LIMIT (WLL)"

NO TERMINOLOGY USED
TO LABEL RATED LOADS

Figure 1-6. Various terminology, or none at all, may be used to describe the safe amount of force that can be applied to a rigging component. Personnel should be aware of the variations and understand that they have the same meaning.

Personnel should note, however, that "breaking strength," also known as "ultimate strength," has a very different meaning. Breaking strength is the force under which the component is expected to fail and must never be used as an allowable sling or equipment load. If ever in doubt regarding the meaning of a similar term for a rigging component, workers should always consult with the manufacturer.

CERTIFICATIONS

The regulations in OSHA 29 CFR 1926 Subpart CC require training and certification for signalpersons and crane operators. Some state and municipal governments and companies also require certification for riggers. In accordance with the ANSI/ASSE A10.42 standard, qualified riggers must also be trained in giving hand signals, since they often have to act as both rigger and signalperson.

A number of third-party organizations provide training in each of these areas. Many organizations also have certification programs for documenting the qualifications of rigging, hoisting, and signaling personnel. Some governmental bodies also administer rigger certification programs and issue licenses. Employers may dictate which organization's certifications meet the needs of their safety program, or they may conduct their own training programs and issue credentials instead of, or in addition to, outside certifications. Periodic refresher training and recertification may also be required.

Exam Preparation

Certification typically requires passing both a written exam and a practical skills test. Some individuals may opt to take the exam soon after the training program, and others may wait weeks or months, if this is an option. The extra time can be spent studying and possibly gaining some on-the-job exposure in the exam topics, such as by assisting a qualified person.

Exam preparation should begin by learning as much as possible about the certification. Students should review the prerequisites and ensure that they qualify for the exam. The certification should be the right one for the student's position and experience. Students should research both the employers who accept this certification and how widely recognized and accepted the certification is compared to others. It may be helpful to speak with individuals who have the same, or similar, certifications.

Next, students should review the outline of the exam, which typically lists the topics covered by the exam, along with the number of questions on each topic. Exams vary depending on the certification organization, but most organizations should provide a general overview of the exam format. Students should note the differences in the written and practical components, the method of testing (for example, paper-and-pencil versus computer), time limits, whether unanswered questions will be counted as wrong, and the passing grades.

The rules for retaking the exam should be reviewed in case of failure. There may be a minimum waiting period before the exam may be retaken. Also, some organizations require the entire exam to be retaken, while other organizations allow students to retake only the sections not passed, such as the practical component.

The certification organization may provide a list of recommended study materials for the examination. Students should use training and study materials to review all tested topics. Students should also study the material over time and complete the review several days before the exam. When the time comes to take the exam, a student following these and other general test-taking strategies should be well prepared.

COMMON APPLICATIONS—
LOADING AND UNLOADING WIRE REELS

A wire reel is a spool or cylindrical object that wire is wrapped around so that the wire can be easily stored or transported. There are several methods that can be used to lift wire reels, including using a metal pipe and sling, a long sling configured in a choker hitch, and a reel lift.

In the first method, a metal pipe or bar can be inserted through the center of the reel. A sling can be used to lift the reel by placing the sling eyes over each end of the pipe and connecting the middle of the sling to a hook to lift the reel. Padding must be used to protect the sling from sharp edges.

In the second method, a long nylon sling can be configured into a choker hitch by wrapping the sling completely around the reel and back through the sling. The end of the sling can then be attached to a hook to lift the reel.

In the final method, a reel lift can be used to lift a wire reel. A reel lift contains a pivoting lifting arm that can be opened after being inserted into a wire reel hole. A hook is attached to the end of the reel lift and the lifting arm is pulled against the inside of the reel to lift the reel. Reel lifts allow wire reels to be lifted from either side or while in a nonrolling position.

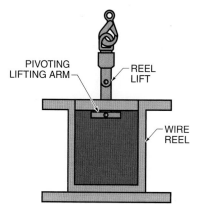

PIVOTING
LIFTING ARM

REEL
LIFT

WIRE
REEL

Chapter 1 Learner Resources

For additional information, visit
qr.njatcdb.org
Item # 1649

Lineworker Rigging Practices

RIGGING EQUIPMENT 2

OBJECTIVES

- Differentiate between the types of hooks and their uses.
- Differentiate between the types of shackles and their uses.
- Differentiate between the types of eyebolts and their uses.
- Describe the use of eyenuts, swivel hoist rings, turnbuckles, links, grips, carabiners, lifting beams, equalizer beams, trolleys, and beam clamps.
- List the general inspection criteria for removing rigging hardware from service.

Rigging equipment includes the devices used to connect lifting slings to the hoist and load. The individual features of the various types of hardware make them each suitable only for certain rigging arrangements. Rigging hardware selection is critical for ensuring a safe and successful lift. Each component in the rigging arrangement must have a sufficient size and strength for its selected application.

HARDWARE ATTACHMENTS

Rigging hardware is the hardware used to connect loads to slings, slings to slings, and slings to hoist hooks. The choice of rigging hardware can be critical, as it helps manage the lifting forces and keep the load under control. Rigging a load with the improper hardware attachments or using the attachments improperly can cause equipment failure or allow the load to shift into an unsafe position.

Some attachments are added to a sling when it is manufactured. In these cases, the working load limit of the entire sling should already account for the strength of the attachments. However, various attachments can be purchased separately and attached individually. A rigging assembly is only as strong as its weakest component, so the strength of each added attachment and hardware must be considered. Attachments must be used for their intended purpose and in accordance with the manufacturer's instructions.

Rigging hardware is designed and manufactured to have sufficient ductility to deform before losing the ability to support the load. Therefore, makeshift hooks, links, or fasteners should not be used, since their strengths are unknown. Hardware that has not been designed for lifting may fail suddenly and catastrophically without deforming or giving visual clues that it is overloaded.

Hooks

A hook is usually the primary link between the rigged load and the hoisting equipment. A *hook* is a curved implement used for quickly and temporarily connecting rigging to loads or lifting equipment. Hooks are made in various designs and sizes. The primary design variations are based on the shape of the hook, method of attachment, and latch arrangement. **See Figure 2-1.** Hooks are available in many combinations of these designs.

HOOKS

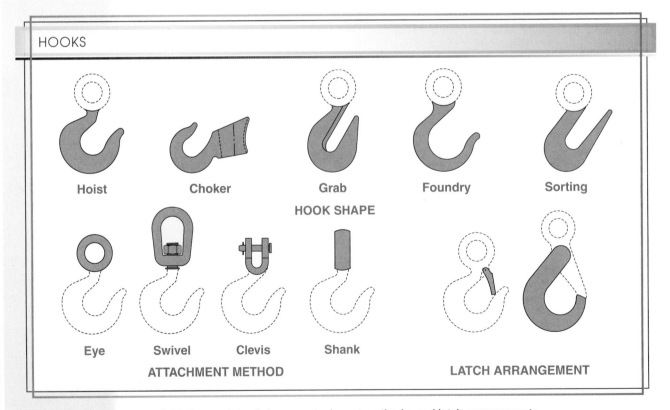

Figure 2-1. Hooks are available in a variety of shapes, attachment methods, and latch arrangements.

The most common hook design is the hoist hook. A *hoist hook* is a hook with a rounded shape that is suitable for most rigging and lifting applications. Other hook shapes are used for special applications. A *choker hook* is a sliding hook used to form a choker sling when hooked to a sling eye. A *grab hook* is a hook that can engage and securely hold a chain link. A grab hook can also be used to shorten a chain sling leg. A *foundry hook* is a hook with a wide, deep throat that fits the handles of molds or casting. A *sorting hook* is a hook with a straight, tapered tip that can be used to directly hold plates, cylinders, and other shapes that allow full engagement.

The top portion of a hook provides a means to attach the hook to a hoist. Other hardware is typically needed to make this attachment to a rope or chain. Also, swivel hooks allow the hook to turn without twisting the hoisting line. However, a regular swivel hook is used only as an aid for positioning the hook and is not intended for load rotation. Special load-rotation swivel hooks equipped with roller bearings are available for such applications.

On hoist and choker hooks, latches are usually required to prevent slings or other lifting hardware from slipping out of the hook. A *latch* is a load-operated jaw, rotating gate, or spring-loaded clip that closes off the opening of a hook. Unless a qualified person determines that a load can be lifted more safely without the use of latches on the hooks, the latches must be present and operational on the hooks of all hoisting equipment. Manufacturer recommendations should always be followed. Care should be taken to ensure that loads are supported by the hook and not by the latch.

Hooks are normally made of forged steel and are usually selected based on the working load limit of the hoist, chain, or other rigging equipment to which the hook is attached. **See Figure 2-2.** Hoist hooks are heat treated with a quenching and tempering process that enhances their impact, ductility, and fatigue properties. For this reason, hooks should not be welded on or subjected to high temperatures, such as being heated by a torch, in order to attempt a repair. Hoist hooks are designed to deform when they are overloaded, rather than breaking unexpectedly, and improper heating or cooling will reduce the strength of the hook and may cause it to fail before deforming. Also, heat can reduce the strength of hooks, so they should never be used at temperatures above 400°F (204°C) without the approval of the manufacturer or a qualified person.

RATED LOADS OF HOIST HOOKS WITH EYE ATTACHMENTS

Throat Opening†	Material	
	Carbon Steel*	Alloy Steel*
0.88	1500	2000
0.97	2000	3000
1.00	3000	4000
1.12	4000	6000
1.36	6000	10,000
1.50	10,000	14,000
1.75	15,000	22,000
1.91	20,000	30,000
2.75	30,000	44,000

* in lb, with a safety factor of 5
† in in.

Figure 2-2. Larger hooks have greater load capacities. However, hooks are often identified by their rated capacities in tons or pounds rather than size.

Hooks must be attached to loads in such a way that the hooks are fully engaged and loaded in-line. The load should be supported from the base of the hook, also called the "bowl" or "saddle," and not from the tip. **See Figure 2-3.** When used at angles other than vertical, the tip of the hook must point outward in order to reduce the chance that the hook will slip off the lift point. Hooks must not be loaded on the side, back, or tip.

PROPER HOOK LOADING

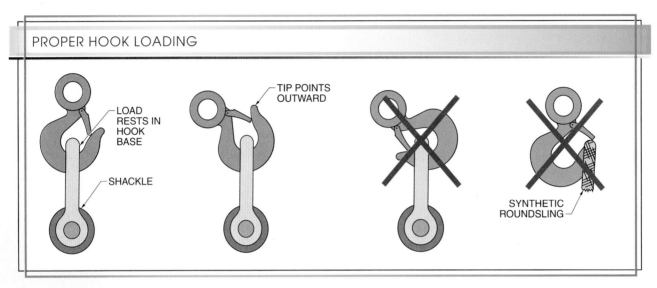

Figure 2-3. Hooks should be arranged so that loads are in the base of the hook, where they are not likely to slip out.

When a hook is used to collect two slings, the maximum included angle is 90°. Some manufacturers include 45° (from vertical) angle marks on the hook to ensure compliance. **See Figure 2-4.** For collecting more than two slings, shackles or master links must be used.

MAXIMUM INCLUDED ANGLE FOR HOOKS

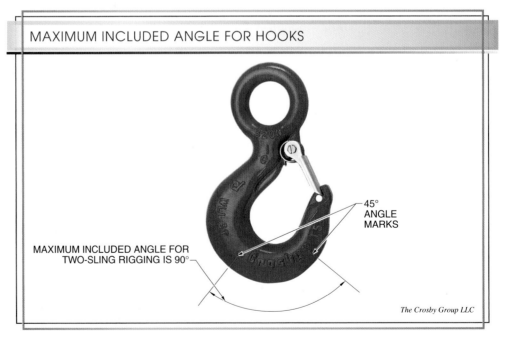

The Crosby Group LLC

Figure 2-4. Hooks typically include 45° (from vertical) angle marks to indicate the maximum included angle for rigging with two slings.

Shackles

A *shackle* is a U-shaped metal connector with holes drilled into the ends for receiving a removable pin or bolt. When assembled, the shackle forms a complete loop that is used to securely connect other rigging equipment. The removal of the pin or bolt opens the loop. A shackle is commonly used to make the connection between the rigging assembly and the hoisting hook, though it can also be used for other purposes. Shackles are made in various shapes and pin designs. **See Figure 2-5.** The shape of the shackle may determine how many and what type of slings (wire rope, web sling, chain, etc.) can be collected.

Screw-pin shackles have threaded pins that are screwed into a threaded hole in the opposite "ear" of the shackle. To be properly used, the screw pin must be fully engaged and the shoulder must be snug against the body of the shackle. These shackles are appropriate for a variety of temporary rigging applications. Long-term use or other in-house safety requirements may require that the pin of a screw-pin shackle be wired, or "moused," to the body of the shackle to prevent the pin from unscrewing.

SHACKLES

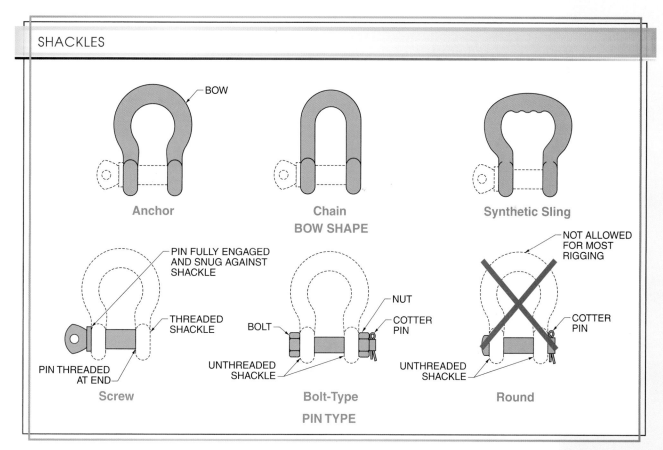

Figure 2-5. The most common shackle shapes are the anchor and chain types. Shackle pin designs include screw pins, bolt-type pins, and round pins, though round pins are not allowed for lifting applications.

Bolt-type shackles use a bolt-like pin that is passed through both ends of the shackle and secured with a nut and cotter pin. Bolt-type shackles are commonly used for a variety of more permanent rigging applications, but they may also be required for all applications in some facilities because of their higher level of safety. The cotter pins should be retained and maintained. Cotter pins should not be substituted with bits of wire, welding rod, or other objects. Round-pin shackles, which use only a cotter pin to secure the main pin, cannot be used for most lifting applications.

The strength of a shackle is determined by its steel composition, manufacturing and heat treatment process, and size. **See Figure 2-6.** While lifting shackles, use an alloy steel pin, the body of the shackle may be made of either carbon steel or alloy steel. An alloy steel shackle is significantly stronger than a comparably sized carbon steel shackle. Shackles are marked with the manufacturer's name or trademark, working load limit, and size. The size is based upon the thickness of the bow and not on the diameter of the pin.

RATED LOADS OF ANCHOR SHACKLES

Nominal Size†	Inside Width at Pin†	Pin Diameter†	Material	
			Carbon Steel*	Alloy Steel*
3/16	0.38	0.25	667	—
1/4	0.47	0.31	1000	—
5/16	0.53	0.38	1500	—
3/8	0.66	0.44	2000	4000
7/16	0.75	0.50	3000	5200
1/2	0.81	0.63	4000	6600
5/8	1.06	0.75	6500	10,000
3/4	1.25	0.88	9500	14,000
7/8	1.44	1.00	13,000	19,000
1	1.69	1.13	17,000	25,000
1 1/8	1.81	1.25	19,000	30,000
1 1/4	2.03	1.38	24,000	36,000
1 3/8	2.25	1.50	27,000	42,000
1 1/2	2.38	1.63	34,000	60,000

* in lb, with a safety factor of 5
† in in.

Figure 2-6. Shackles are identified by their size, inside width, and pin diameter. Larger shackles provide greater load-lifting capacity.

The orientation of a shackle in use is important. Side loading should be avoided. For example, when a shackle is used to connect two other pieces of rigging equipment, one connection must be on the pin, and the other must be in the base of the bow. Washers may need to be placed on both sides of the shackle pin in order to maintain alignment. The shackle must not be loaded with these two forces on the sides of the bow, which will tend to pull the bow open and distort the shackle.

In some cases, the shackle may be side-loaded. For example, shackles may be attached to a lift point and pulled from the side. However, the working load limit of the shackle will be reduced by up to 50%, and the direction of force cannot exceed 90° from the in-line position. **See Figure 2-7.**

FACToid

Complex lifts and some common lifts require the development of a lift plan. A lift plan is an evaluation of the potential hazards of a lift and the equipment and procedures required to safely execute the lift. A lift plan is used by engineering, safety, field, and any other necessary personnel to prepare for a lift.

SHACKLES LOADED AT AN ANGLE

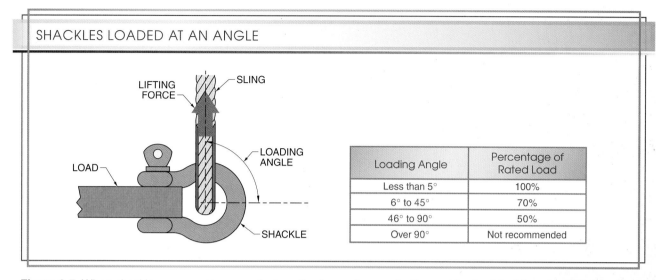

Loading Angle	Percentage of Rated Load
Less than 5°	100%
6° to 45°	70%
46° to 90°	50%
Over 90°	Not recommended

Figure 2-7. When shackles are used in a way that applies force at an angle, the working load limit is reduced.

When multiple rigging slings are collected in a shackle, they should all bear on the bow, with the lift hook centered on the pin. The included angle between the slings must not exceed 120° in order to prevent excessive side loading.

When using shackles as part of a rigging project, some general rules should be followed. Shackles should never be overloaded. Shock loading of the shackle and contact with sharp edges should be avoided because the shackle could be damaged. Shackles should be oriented so that the screw pin will not unscrew as the load is lifted. When using a shackle to form a choker with a bridle sling, the shackle pin should be in the eye of the sling and not against the body of the sling. Common bolts should never be substituted for the alloy pins of a shackle.

OSHA REGULATIONS AND LINEMAN SAFETY

1926.502(d)(6)(i-v)
Unless the snaphook is a locking type and designed for the following connections, snaphooks shall not be engaged: direclty to webbing, rope or wire rope; to each other; to a dee-ring to which another snaphook or other connector is attached; to a horizontal lifeline; or to any object which is incompatibly shaped or dimensioned in relation to the snaphook such that unintentional disengagement could occur by the connnected object being able to depress the snaphook keeper and release itself.

1926.502(d)(15)(i-ii)
Anchorages used for attachment of personal fall arrest equipment shall be independent of any anchorage being used to support or suspend platforms and capable of supporting at least 5,000 pounds (22.2 kN) per employee attached, or shall be designed, installed, and used as follows: as part of a complete personal fall arrest system which maintains a safety factor of at least two; and under the supervision of a qualified person.

1926.502(d)(21)
Personal fall arrest systems shall be inspected prior to each use for wear, damage and other deterioration, and defective components shall be removed from service.

Eyebolts

Some loads are preequipped by the manufacturer with lift points, which simplifies balancing, rigging, and lifting. The lift points may or may not already include fasteners, such as eyebolts. An *eyebolt* is a bolt with a looped head that is fastened to a load to provide a lift point. **See Figure 2-8.** Eyebolts are attached by screwing them into a threaded hole or inserting an eyebolt shank through a hole and securing the eyebolt with a nut. Eyebolts are also known as lifting eyes.

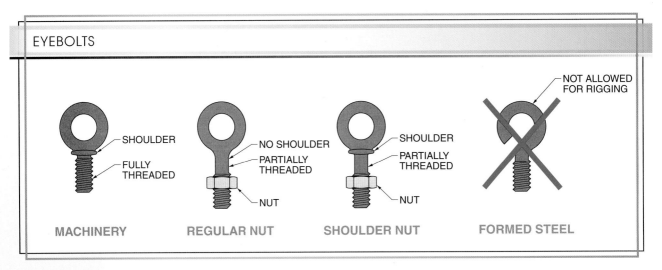

EYEBOLTS

MACHINERY — SHOULDER / FULLY THREADED

REGULAR NUT — NO SHOULDER / PARTIALLY THREADED / NUT

SHOULDER NUT — SHOULDER / PARTIALLY THREADED / NUT

FORMED STEEL — NOT ALLOWED FOR RIGGING

Figure 2-8. Forged machinery or nut eyebolts are used in various rigging and lifting applications.

The load capacity of an eyebolt is greatest when it is pulled vertically, in a direction directly opposite to the shank. Eyebolts with a shoulder may be loaded at other angles, but only within the plane of the loop. Also, when forces are angular, the effective capacity of the eyebolt is reduced. **See Figure 2-9.** Specific working load limit information must be obtained for a particular eyebolt because they can vary significantly between manufacturers. However, general working load limits of eyebolts may be used as a primary estimation when designing lifting assemblies. **See Figure 2-10.**

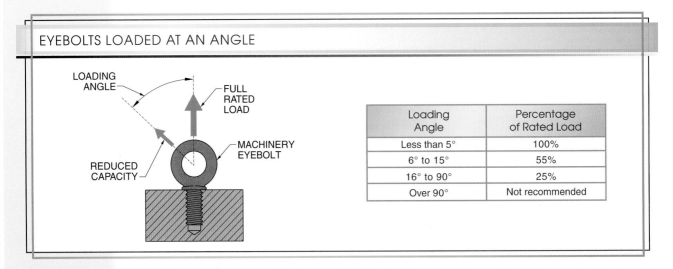

EYEBOLTS LOADED AT AN ANGLE

LOADING ANGLE — FULL RATED LOAD — MACHINERY EYEBOLT — REDUCED CAPACITY

Loading Angle	Percentage of Rated Load
Less than 5°	100%
6° to 15°	55%
16° to 90°	25%
Over 90°	Not recommended

Figure 2-9. Eyebolts have the greatest load capacity when used with in-line forces. Their capacity at other angles is reduced.

RATED LOADS OF SHOULDERED EYEBOLTS

Eyebolt Size†	Angle of Pull*			
	0°	30°	45°	60°
¼	650	420	195	160
5/16	1200	780	360	300
⅜	1550	1000	465	380
½	2600	1690	780	650
⅝	5200	3380	1560	1300
¾	7200	4680	2160	1800
⅞	10,600	6890	3180	2650
1	13,300	8645	3990	3325
1¼	21,000	13,650	6300	5250
1½	24,000	15,600	7200	6000

* in lb
† in in.

Figure 2-10. Working load limits at various angles are sometimes available from manufacturer tables.

The loop of an eyebolt may be formed or forged. Formed loops are rods that are bent around to form a loop. Because one end is open and the steel was deformed to make the loop, these loops are not strong enough for lifting applications and should not be used. Forged loops are pressed into their closed-loop shape in a process that keeps the steel strong. Only forged steel eyebolts should be used for lifting applications. The common types of forged eyebolts are machinery and nut eyebolts.

Machinery Eyebolts. The machinery eyebolt is the most commonly used type of eyebolt. Machinery eyebolts have a shoulder and a fully threaded shank. A machinery eyebolt must be screwed in until the shoulder is tight against the load. The threaded hole must be deep enough to ensure that the threads of the eyebolt are engaged by at least 1½ thread diameters. **See Figure 2-11.** A shim or washer may be used to create a tight shoulder-to-load fit if the threaded hole is too short but only up to the thickness of one thread pitch. This allows the eyebolt to be tightened against rotation during the lift.

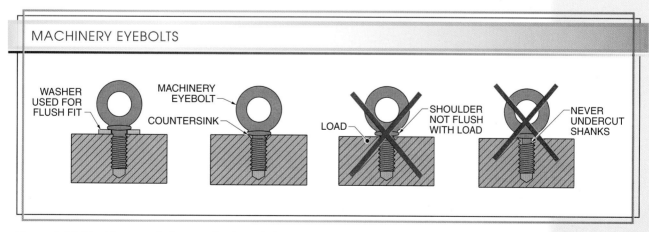

MACHINERY EYEBOLTS

WASHER USED FOR FLUSH FIT

MACHINERY EYEBOLT

COUNTERSINK

LOAD

SHOULDER NOT FLUSH WITH LOAD

NEVER UNDERCUT SHANKS

Figure 2-11. Machinery eyebolts must be threaded into a load until the shoulder is tightly against the load surface.

A shim or washer can also be used to adjust the eyebolt orientation so that an angular lifting force remains in the plane of the eye. **See Figure 2-12.** With the proper shim thickness, the eyebolt will be in proper alignment after being fully tightened. The shim thickness is determined by the size of the eyebolt and how much it must be turned (unthreaded) to be aligned properly. Also, the inside diameter of the shim or washer must be matched to the size of the eyebolt.

EYEBOLT ALIGNMENT FOR ANGULAR FORCES

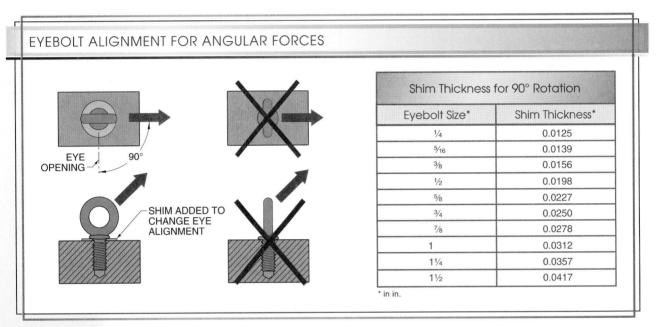

Shim Thickness for 90° Rotation	
Eyebolt Size*	Shim Thickness*
¼	0.0125
5/16	0.0139
3/8	0.0156
½	0.0198
5/8	0.0227
¾	0.0250
7/8	0.0278
1	0.0312
1¼	0.0357
1½	0.0417

* in in.

Figure 2-12. When used in rigging that applies angular forces, eyebolts must be turned until the loop is in-line with the rigging. Shims can be used to achieve the proper orientation while the eyebolt shoulder is snug up against the load.

Nut Eyebolts. Nut eyebolts are not fully threaded and require nuts to be secured in place. These eyebolts are often used when the thinness of the load material prevents the proper use of machinery eyebolts. In these cases, nut eyebolts are installed into either threaded or unthreaded through holes. Nut eyebolts include regular nut eyebolts and shoulder nut eyebolts.

Regular nut eyebolts are not permitted for rigging in some jurisdictions due to their lack of a shoulder. Often, they are only permitted for special circumstances and vertical pulls. For a threaded through hole in material that has a thickness greater than one eyebolt shank diameter, a nut secures the attachment from the bolt end. **See Figure 2-13.** For thinner materials, a nut is required in both sides. For an unthreaded through hole, two nuts are firmly tightened against each other on the bolt end with a third nut on the loop side.

FACToid

When using eyebolts and slings for rigging, slings should not be run through a pair of eyebolts because this will reduce the effective angle of lift and place more stress on the eyebolts. Slings should not be forced through the eyebolts because the load and the angle of loading may be altered. Eyebolts should not be painted, since the paint could cover up possible flaws. Also, the point of a hook should not be inserted into an eyebolt. A shackle should be used instead.

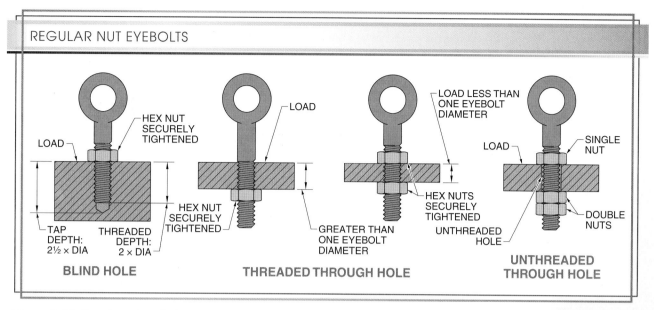

REGULAR NUT EYEBOLTS

Figure 2-13. Regular nut eyebolts use one to three nuts to secure the eyebolt against the load.

When a shoulder nut eyebolt is used, the threaded portion of the shank must protrude through the load enough to allow full engagement of the nut. **See Figure 2-14.** Washers are used when the unthreaded portion of the shoulder nut eyebolt protrudes so far that the nut cannot be tightened securely against the load. The washers should only be installed on the nut side of the eyebolt and not between the shoulder and the load. Shoulder nut eyebolts may be loaded at angles other than in-line, though there will be a corresponding reduction in load capacity.

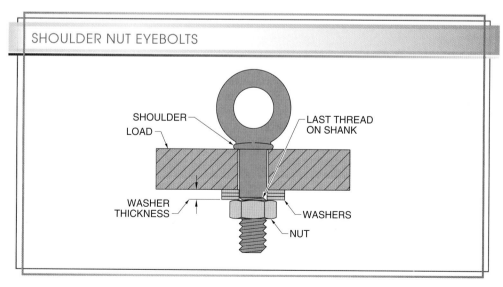

SHOULDER NUT EYEBOLTS

Figure 2-14. Shoulder nut eyebolts have a shoulder that can be placed tightly against the load surface and require a nut to secure the opposite end.

Eyenuts

Eyenuts are another way to provide convenient rigging attachments to load lift points. An *eyenut* is a loop-shaped nut that is fastened to a load to provide a lift point. **See Figure 2-15.** Eyenuts are only used with through holes. A bolt is inserted into the through hole and fastened with a fully engaged eyenut. A separate regular hex nut may also be used to further secure the eyenut. The resulting arrangement is similar to an eyebolt. Eyenuts are suitable only for vertical lifting forces.

EYENUTS

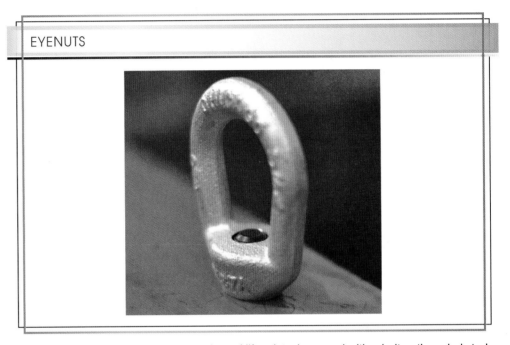

Figure 2-15. Eyenuts provide a loop-shaped lift point when used with a bolt or threaded stud.

Swivel Hoist Rings

Swivel hoist rings are similar to eyebolts, except that they can rotate under load without unfastening. A *swivel hoist ring* is a freely rotating attachment that is fastened to a load to provide a lift point. **See Figure 2-16.** This allows the swivel to rotate to remain in alignment with the lifting slings. It also prevents the side loading that might occur if the eyebolts were improperly aligned or if the attachment hardware were to snag or bind on fixed eyebolts. Shims or washers must not be used between the base of the swivel and the load, and the mounting bolt must be torqued in accordance with the manufacturer's recommendation.

FACToid

All rigging gear and equipment provided by an employer must be inspected before each shift and at intervals during use to minimize the possibility of a rigging failure. The person authorized and qualified to rig must always pay close attention to details. One careless act can result in serious injury or death and property damage. The rigging capacity and the material to be lifted must match. Improper rigging of a load or a rigging failure can expose riggers and other workers nearby to a variety of potential hazards. Riggers have been injured or killed when loads have slipped from the rigging or when the rigging has failed.

SWIVEL HOIST RINGS

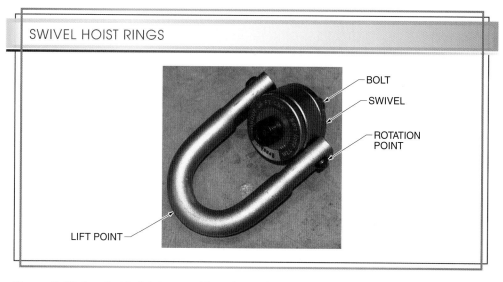

BOLT
SWIVEL
ROTATION POINT
LIFT POINT

Figure 2-16. A swivel hoist ring provides a loop-shaped lift point that can swivel and rotate.

Turnbuckles

A *turnbuckle* is an adjustable-length attachment used for connecting other rigging components. A turnbuckle consists of a pair of threaded fittings screwed into a metal body. **See Figure 2-17.** The fittings are commonly two eyebolts, though they may also be any combination of eyebolts, hooks, or pinned jaws. The body is an open or pipe-type frame.

TURNBUCKLES

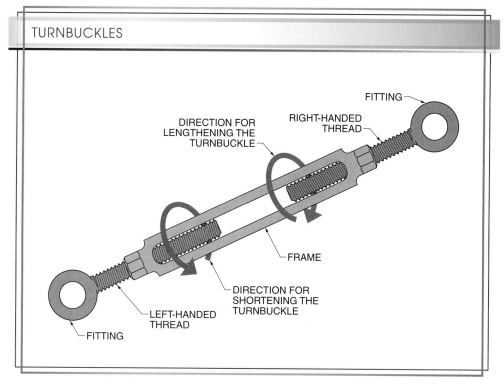

FITTING
RIGHT-HANDED THREAD
DIRECTION FOR LENGTHENING THE TURNBUCKLE
FRAME
DIRECTION FOR SHORTENING THE TURNBUCKLE
LEFT-HANDED THREAD
FITTING

Figure 2-17. A turnbuckle is a pair of threaded fittings screwed into the opposite ends of a frame body and can be adjusted for length by twisting the frame in relation to the fittings.

One fitting has a right-hand thread and the other has a left-hand thread. Due to this arrangement, when the ends are constrained, the overall length of the turnbuckle can be adjusted by rotating the body. Rotating one direction lengthens the turnbuckle, while rotating the other direction shortens it. A common open-ended wrench is used to adjust a turnbuckle by applying it to the flats on the body.

Turnbuckles are typically used in complex rigging applications in which an asymmetrical load requires slings of different lengths. Turnbuckles are used to achieve exactly the right length to keep the load level and stable. **See Figure 2-18.** The length that the turnbuckle is adjusted depends on its size, though it is generally short compared to the length of the rigging. Therefore, slings should be chosen as close as possible to the desired length while also accounting for the overall length of the turnbuckle, which is then used to make fine length adjustments. The full engagement of the threads between the fittings and the body of the turnbuckle must be ensured in order for safe operation.

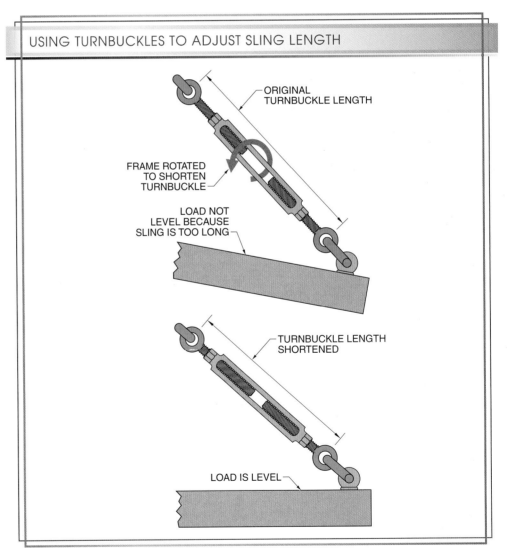

USING TURNBUCKLES TO ADJUST SLING LENGTH

ORIGINAL TURNBUCKLE LENGTH

FRAME ROTATED TO SHORTEN TURNBUCKLE

LOAD NOT LEVEL BECAUSE SLING IS TOO LONG

TURNBUCKLE LENGTH SHORTENED

LOAD IS LEVEL

Figure 2-18. Turnbuckles are used to finely adjust sling length in rigging applications, which is most commonly needed for complex rigging of asymmetrical loads.

Turnbuckles are designed only for in-line forces. Nothing should be attached to the body that pulls on the turnbuckle from the side. When installed for long-term use or when vibration may loosen the turnbuckle, the body may be secured against rotating through the use of pins, safety wires, or nuts. **See Figure 2-19.** If nuts are used, they should be tightened only enough to prevent rotation of the barrel. Overtightening the nuts prestresses the turnbuckle and may severely reduce its load capacity.

SECURING TURNBUCKLES

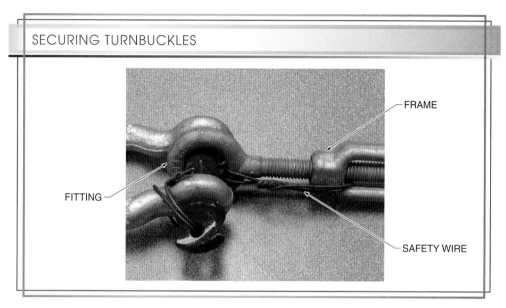

Figure 2-19. Turnbuckles may be secured against unintentional twisting by locking the position of the frame in relation to the end fittings.

Links

A *link* is a plain, rigid loop attached to the end of a sling or hung on a crane hook that helps connect multiple rigging components together. Links provide a larger connection for joining multiple slings or shackles where the opening of a hook is limited and would cause binding of the slings or improper loading of the hook. Links are often permanently installed to slings by the sling manufacturer. Links are commonly used as the hoist connections for multiple-leg sling assemblies. **See Figure 2-20.** However, links can also be used on the load-attachment end of slings.

PERMANENTLY INSTALLED LINKS

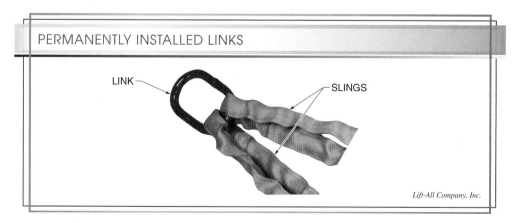

Lift-All Company, Inc.

Figure 2-20. Links are often permanently installed by a manufacturer at the ends of two or more slings.

Links are available in different shapes. **See Figure 2-21.** An end link resembles a chain link. There are also ring and pear-shaped links. Each shape is suited for certain rigging combinations. For example, a pear-shaped link is suited for use with a hook at the top and multiple slings along the bottom. Link working load limits are based on a single vertical load or a collection of sling legs with an included angle of 120° or less. **See Figure 2-22.**

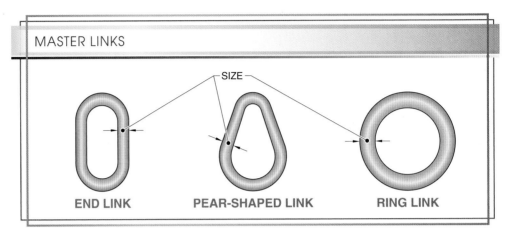

MASTER LINKS

SIZE

END LINK   PEAR-SHAPED LINK   RING LINK

Figure 2-21. Master links are available in different sizes and shapes for collecting multiple slings and connecting rigging to lifting equipment.

RATED LOADS OF MASTER LINKS

Nominal Size†	Material			
	End	Ring	Pear Shape	
	Alloy Steel*	Carbon Steel*	Alloy Steel*	Carbon Steel*
½	7000	—	7000	2900
⅝	9000	—	9000	4200
¾	12,300	—	12,300	6000
⅞	15,000	7200	15,000	8300
1	24,360	10,800	24,360	10,800
1⅛	—	10,400	30,600	—
1¼	36,000	17,000	36,000	16,750
1⅜	—	19,000	43,000	20,500
1½	54,300	—	54,300	—
1⅝	—	—	62,600	—
1¾	84,900	—	84,900	—
2	102,600	—	102,600	—

* in lb, with a safety factor of 5
† in in.

Figure 2-22. Rated capacities for master links are based on their size, shape, and material.

Grips

A *grip* is a mechanical device used for pulling cable, wire, or rope. Linemen use grips to help tension wire and cable during installation and while making repairs. The greater the force applied to the pulling point on the grip, the greater the grip pressure on the cable or wire. **See Figure 2-23.**

GRIPS

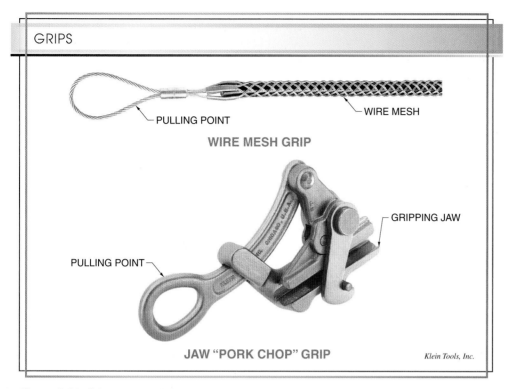

Figure 2-23. Grips are used to pull cable or conductors.

Carabiners

A *carabiner* is an oblong metal ring with one spring-hinged side that is used in climbing as a connector and to hold a freely running rope. Carabiners are typically used by linemen for fall protection and are increasingly being used for material handling and rigging. **See Figure 2-24.**

CARABINERS

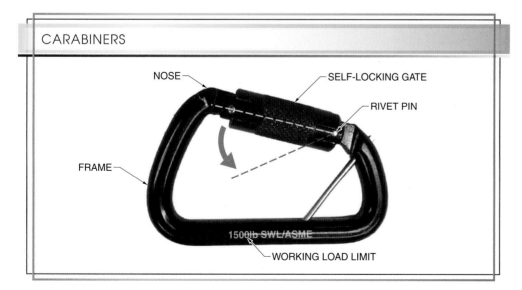

Figure 2-24. Approved carabiners that are listed for material handling are used by linemen for material hoisting and rigging.

Only approved carabiners may be used for material hoisting and rigging. These carabiners are not suitable for fall protection purposes. They must be labeled for material use only and include the appropriate working load limit.

Lifting Beams

A *lifting beam,* or spreader beam, is a bar, beam, or tube used to change the direction of rigging forces at the load from angled to vertical. This is particularly important if angled lifting forces will cause a flexible load to buckle or are otherwise not tolerated by the attachment points. **See Figure 2-25.**

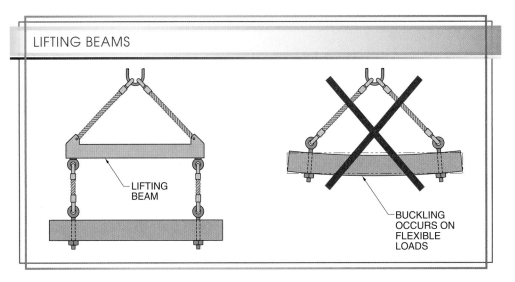

LIFTING BEAMS

LIFTING BEAM

BUCKLING OCCURS ON FLEXIBLE LOADS

Figure 2-25. If angular forces could cause a flexible load to buckle, a lifting beam can be added to the rigging to change the direction of the forces on the load.

Lifting beams are generally constructed of steel, but they may be fabricated out of other materials for special applications. For example, fiberglass lifting beams may be used in corrosive environments. Because the lifting beam is a "below the hook" lifting device, its weight must be considered when calculating the load on the rigging above it.

The lift points on lifting beams may be fixed or adjustable. Also, multiple lift points may be provided along the beam so that the proper sling angles can be achieved.

Regardless of the type, the lifting beam must be designed and constructed under the supervision of a qualified person. After construction, a load test of 125% of the working load limit must be applied to the beam. Also, any operable components on the beam must be tested for functionality.

Equalizer Beams

An *equalizer beam* is a load leveler that is used to stabilize a load by distributing the load equally between two sling legs or equalizing loads on two hoist lines when performing a tandem lift. Equalizer beams are helpful when performing heavy lifts that would require two cranes to distribute the weight of the load between the two cranes so that they can handle the lift. **See Figure 2-26.**

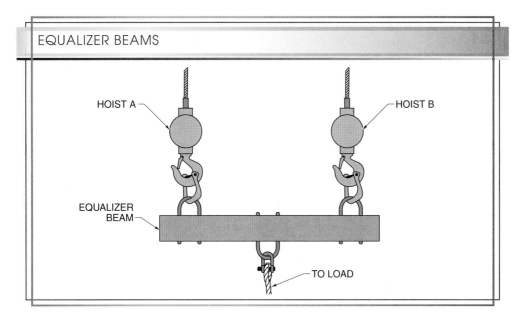

EQUALIZER BEAMS

HOIST A

HOIST B

EQUALIZER BEAM

TO LOAD

Figure 2-26. Equalizer beams are used to stabilize loads by distributing the load equally on two hoist lines when performing a tandem lift.

Lever-Operated Hoists

A *lever-operated hoist* is a manually operated hoist that uses a lever to provide torque to a gear drive. Most lever-operated hoists use ratchets to transfer power and prevent reverse rotation. A ratchet has a toothed wheel that is prevented from rotating backward by engaging with a spring-loaded pawl. **See Figure 2-27.** The ratchet wheel advances with each stroke of the lever, either directly or through a gear drive, to provide additional mechanical advantage. A release mechanism reduces tension on the hoisting line in a controlled manner. Lineworkers typically use strap, chain, and model BB lever-operated hoists.

Trolleys

A *trolley* is a wheeled assembly that travels horizontally along a crane beam or boom. **See Figure 2-28.** The wheels on a trolley allow loads to be hoisted and then moved laterally. Trolleys are typically used in conjunction with I-beams or A-frames. Trolleys must be adjusted when installed so that the wheel width is ⅛″ greater than the flange that the trolley is riding on.

Trolleys are typically used to support lever- or chain-type hoists, and care should be taken that none of the components used have working load limits higher than the equipment above them. If a trolley has a working load limit higher than the A-frame above it, or if the hoist has a rated capacity greater than the trolley, the entire assembly must be reduced to the rating of the lowest rated component.

Beam Clamps

A *beam clamp* is a fixed device for attaching a hoist to the bottom of a beam at a single location. Multiple beam clamps and hoists may be used to drift a load along the length of a beam. As a rule, the angle of the hoisting line when using a beam clamp in a drifting operation should not exceed 15° from vertical. This is because excessive horizontal loading of a beam clamp can bend and break the clamp. As with trolleys, the working load limit of the beam clamp should also be coordinated with the other rigging equipment being used.

LEVER-OPERATED HOISTS

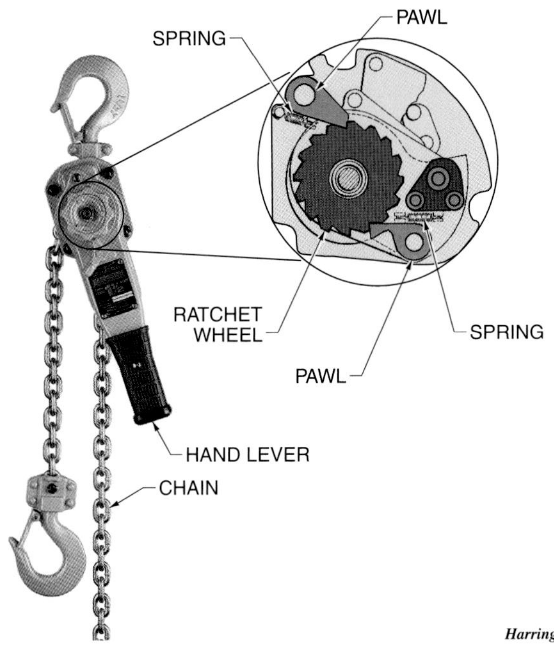

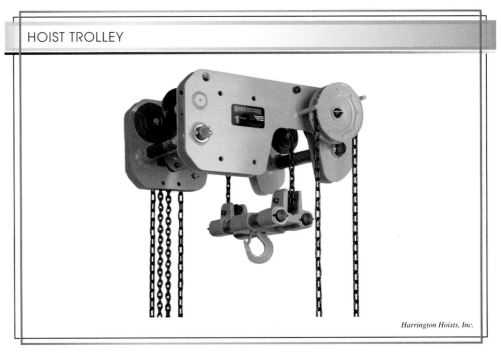

SPRING

PAWL

RATCHET
WHEEL

SPRING

PAWL

HAND LEVER

CHAIN

Harrington Hoists, Inc.

Figure 2-27. A lever-operated hoist uses a ratchet to convert the back-and-forth motion of a lever to torque.

HOIST TROLLEY

Harrington Hoists, Inc.

Figure 2-28. It is important that trolley wheels are close enough to provide a secure attachment to the beam but not so close that they impede free travel.

HARDWARE INSPECTION

Normal use will cause minor scratches, missing paint or coatings, and worn surfaces on rigging hardware. These types of minor damage are typically not cause for concern. However, significant damage and distortions in shape are evidence of misuse, excessive loading, or exposure to damaging conditions. Prior to use, all rigging hardware should be inspected by a person trained to identify significant damage. Written inspection records are not required for hardware.

All types of hardware should be inspected for visible signs of damage and removed from service if required. These include, but are not limited to, the following signs:

- missing or unreadable manufacturer name or working load limit information

- evidence of heat damage, such as weld splatter

- excessive pitting or corrosion

- bent, twisted, or elongated parts

- gouged, cracked, or broken parts

- excessive thread damage

- reduction in any specified dimension of 10% or more

- evidence of unauthorized modifications, such as drilling, machining, or grinding

- for hooks, expansion of the throat opening by at least 5% or ¼″

- for hooks, missing or inoperable latch mechanisms

- for hooks, damage to hook attachments

- for shackles, incomplete pin engagements

- for swivel hoist rings, lack of ability to freely rotate or pivot

- any other conditions that cause doubt about the hardware's safety

Some types of damage on some hardware can be repaired but only by a qualified person. Information on serviceable parts and repair techniques is typically available from the manufacturer. However, it is usually more cost-effective to simply replace hardware with new items.

RIGGING EQUIPMENT INSPECTION

All rigging slings, attachments, hardware, and fittings must be regularly cleaned and individually inspected. Inspection is a critical part of maintaining rigging equipment. Damaged equipment must be dealt with promptly to avoid not only safety hazards but also operations downtime resulting from a lack of proper equipment being available.

Inspection Types

The use of standardized inspections, conducted by competent persons, is necessary for maintaining an inventory of safe and appropriate rigging equipment. All rigging equipment must undergo three types of inspections: initial inspection before being placed in service, periodic inspection at certain intervals, and frequent inspection before each use. **See Figure 2-29.**

Initial Inspections. An *initial inspection* is a rigging equipment inspection performed before equipment is placed into service in which a designated person ensures that the equipment meets the applicable government regulations and industry standards. It also must meet the required specifications for the loads and lifting conditions of its expected use.

RIGGING EQUIPMENT INSPECTION TYPES

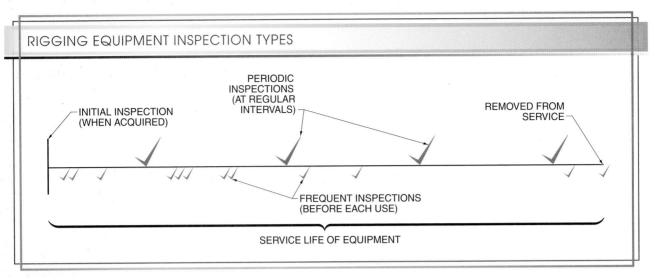

Figure 2-29. The three types of rigging equipment inspections, initial inspection, periodic inspection, and frequent inspection, are performed at different times during the useful service life of the equipment.

Equipment that does not meet the initial inspection requirements may have a manufacturing defect, an improper label, or shipping damage. Alternatively, the equipment may be in good condition but not appropriate for the application needed, due to either ordering or fulfillment error. In all of these cases, the equipment manufacturer should be consulted for advice on returning the equipment and obtaining a suitable replacement.

Periodic Inspections. A *periodic inspection* is a thorough rigging equipment inspection performed at regular intervals. The recommended interval depends on the type of equipment and the conditions of use. Periodic inspections are performed on at least an annual basis by a designated person under normal service conditions. Inspections should be performed on a monthly to quarterly basis if the conditions of use are severe, and even more frequently if a qualified person determines that special service use requires additional inspections. Inspections must be timed so that equipment is removed from service before its condition poses a hazard with continued operation.

Periodic inspections are performed even if the equipment has not been used during the inspection interval. Besides improper use, poor storage conditions can also cause damage, such as corrosion or mold. Maintaining the periodic inspection schedule even when the rigging equipment is not regularly used helps ensure that proper equipment is available when it is needed. In particular, damaged equipment can be replaced or repaired without the added pressure of immediate need as a time constraint.

Frequent Inspections. A *frequent inspection* is a rigging equipment inspection performed at the beginning of each workday or shift by the user of the equipment, or by another designated person. This inspection is less thorough than the other types, but it is critical to ensure that damaged equipment is not inadvertently used. It may be advisable to have a few spares of certain equipment types available at a critical rigging and hoisting job in case some equipment is determined to be unusable during this inspection.

Inspection Checklists

All types of inspections are intended to identify critical defects or damage to the equipment, including damage to its identification or specification tags. Some signs of damage are quantified in order to determine whether the damage is great enough to necessitate removing the equipment from service. For example, scratches on metal fittings are generally not a problem, but deep gouges may be a problem. Therefore, in addition to visual examinations, inspection activities may also include a few types of measurements. **See Figure 2-30.** Simple measuring tools for angle, length, and depth may be required for determining whether the equipment remains within allowable specifications.

INSPECTION MEASUREMENTS

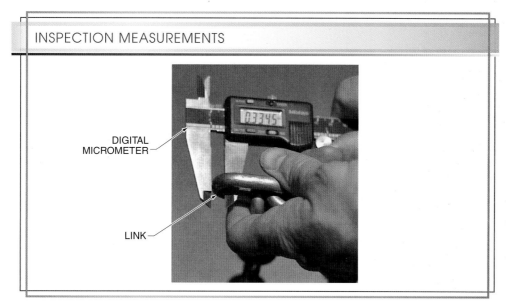

DIGITAL MICROMETER

LINK

Figure 2-30. Some inspections may require the use of measuring tools to determine whether the equipment remains within allowable specifications.

The exact list of inspection criteria depends on the type of equipment and other factors. Applicable regulations and standards specify many minimum inspection criteria. Industry and company rules may impose additional requirements. However, there are many criteria common to all inspection checklists. Generally, all rigging hardware should be inspected for the following deficiencies as part of frequent or periodic inspections:

- missing or unreadable manufacturer's identification
- missing or unreadable working load limit
- deformation from twisting or bending
- elongation from overloading
- misaligned, incomplete, or improper engagement of pins
- excessive thread damage
- hook throat opening increased by more than 5% or ¼″
- malfunctioning or inoperative hook latches
- signs of exposure to excessive heat, welding spatter, or having been welded on
- excessive corrosion or pitting
- cracks, nicks, or gouges
- wear exceeding 10% of the original cross section dimensions
- evidence of modification by improper machining or grinding
- inoperative swivels
- signs of fatigue in sections of rope or slings with frequent loading and bending, such as over pulleys or around loads
- damage to rope or sling ends that terminate at fittings

Damaged Equipment

If damage is found, the component should be immediately removed from service, stored separately, and marked as defective to warn against use. **See Figure 2-31.** In most cases, the equipment cannot be appropriately repaired and must therefore be promptly destroyed and replaced. Given the substantial requirements of proper repair, it is likely that replacement is often the more economical choice. However, when repairs are feasible, they must be performed by a qualified person using the guidelines of the manufacturer or the appropriate ASME standard. Some equipment can be returned to the manufacturer for repair. Unqualified personnel should never attempt repairs.

TAGGING DAMAGED EQUIPMENT

Figure 2-31. To prevent its inadvertent use, damaged equipment must be promptly tagged as such until it can be destroyed or repaired.

If signs of damage are discovered that do not yet exceed allowances, the equipment is permitted to remain in service. However, it may be necessary to increase the frequency of subsequent inspections in order to monitor any changes in the potential problem. This requires careful recordkeeping to identify and track individual pieces of equipment.

RIGGING EQUIPMENT RECORDKEEPING

Written records of the use and inspection of all rigging equipment should be created for each new item. A serial number is issued to and marked permanently on each component, which is used to identify it on each of its inspection records. The inspection records may be kept on paper or electronically. However, either way the inspection records are kept, they should be neat and organized.

The types of information and level of detail in the inspection records may vary depending on the applicable requirements and the amount of rigging done by the user. The minimum recommended information on an inspection document includes the serial number, manufacturer, size, working load limit, attachments, condition notations for each component, and whether the component must be removed from service. **See Figure 2-32.** Additional information typically includes the date, department, inspection type, rigging type, and the name of the inspector. Records for frequent inspections may also include the dates, hours, and locations of each use, along with the lifting conditions.

RIGGING INSPECTION RECORD

Rigging Inspection

Date ___Jan. 14th___   Department ___Shipping___

Inspector ___J. Smith___   Inspection Type ___Monthly___

Rigging: Wire Rope / Fiber Rope / (Web Slings) / Round Slings / Chain / Other _____

Serial Number	Manufacturer	Size	Rated Capacity	Attachments	Condition	Removed from Service?
W-010	NB Slings	6"	9600	triangles	Good	N
W-029	NB Slings	4"	6400	none	½" edge tear	YES— destroyed
W-007	Lift-Tech	6"	16,800	choker/triangle	light wear	N
W-013	Lift-Tech	8"	19,200	none	unreadable tag	YES— returned to mfr.

Figure 2-32. Inspection documents are often used to organize information about rigging equipment inspections.

RIGGING EQUIPMENT STORAGE

Proper care, use, and storage of rigging equipment helps prevent damage and increase safety. Rigging equipment should be kept in a designated and organized area that is clean, dry, and away from harmful fumes or heat. Also, organized storage that keeps equipment sorted by type and size helps riggers locate necessary equipment quickly.

Equipment Cleaning

All rigging equipment should be clean and dry before being stored. However, regular use often exposes the equipment to moisture and contaminants. Since rigging equipment is often used outdoors, it is commonly exposed to dirt and water while in service. Contamination with grease and oil lubricants from either hoists or loads is also a common problem for both indoor and outdoor service. Exposure to certain other chemicals may not be cleanable and will render equipment permanently damaged.

Dirt should be wiped or brushed off, although it may be necessary to wash equipment with clean water or solvent to remove embedded or excessive amounts of dirt. Dirt is particularly likely to get into the weave of synthetic fiber rope and slings. The abrasive action of the dirt wears on fibers and can weaken the rope or sling. If the accumulation becomes excessive, it may be necessary to wash the equipment using a pressure washer, if permitted by the manufacturer's cleaning guidelines. Removing dirt and grit from mating or moving parts, such as pulley bearings, threaded pins, and hook latches, may require disassembly and reassembly by a qualified person. Grit in these components may cause the parts to seize or even permanently damage the mating surfaces when they are used again.

Wet equipment must be thoroughly dried in order to avoid damage from corrosion, mold, mildew, or rot. Hardware and attachments may only require a wipe-down with clean cloths, but rope and synthetic slings require gradual drying in fresh, circulating air before storage.

Rope Storage

Rope should be stored on spools or rolled into coils. Coiling rope usually requires twisting as it is looped so it lies flat. When uncoiling, the rope must be untwisted. **See Figure 2-33.** If the rope is pulled from the coil without untwisting, it will not lay flat and may kink if the loops are drawn tight.

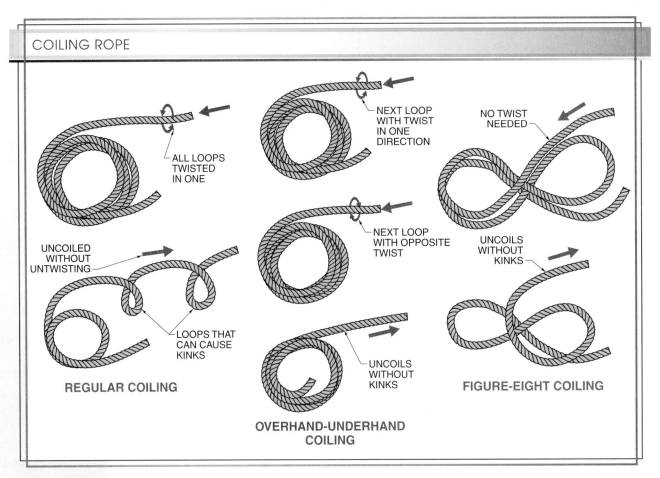

COILING ROPE

NEXT LOOP WITH TWIST IN ONE DIRECTION

ALL LOOPS TWISTED IN ONE

NO TWIST NEEDED

UNCOILED WITHOUT UNTWISTING

NEXT LOOP WITH OPPOSITE TWIST

UNCOILS WITHOUT KINKS

LOOPS THAT CAN CAUSE KINKS

UNCOILS WITHOUT KINKS

REGULAR COILING

OVERHAND-UNDERHAND COILING

FIGURE-EIGHT COILING

Figure 2-33. Regular coiling and uncoiling of rope require significant twisting and untwisting the rope to avoid kinks. Alternatively, the overhand-underhand and figure-eight coiling methods can be used to avoid kinks and make uncoiling easier.

There are two coiling techniques for avoiding this problem. (These techniques can also be used with electrical cords.) First, the overhand-underhand technique adds twists to the rope in opposite directions for each successive loop. This balances the overall twist in the rope so that it will uncoil smoothly. Alternatively, coiling the rope into a figure eight avoids the need to twist the rope. This also helps avoid tangles when the rope must be uncoiled from the starting end. This pattern takes up a lot of space, however, so it is typically suitable only for temporary storage.

Once a rope has been coiled, it usually needs to be secured in some way to maintain the coil. A simple way to do this is to tie a buntline coil. This is formed by taking the last few feet of the rope and wrapping it around the coil four or five times. **See Figure 2-34.** A loop from the working end of the rope is pulled through the coil and then over the top of the coil to secure it. To release, the loop is flipped back over the top of the coil, pulled back through, and unwound. Once coiled, a rope should be stored in a dry location, out of direct sunlight.

BUNTLINE COIL

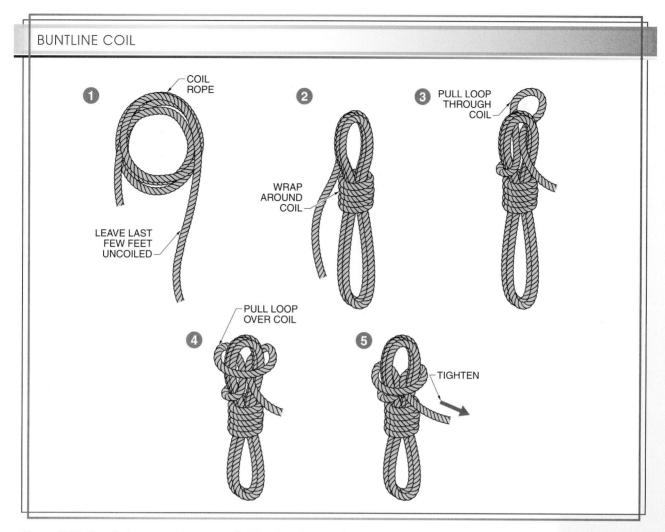

1 COIL ROPE

LEAVE LAST FEW FEET UNCOILED

2 WRAP AROUND COIL

3 PULL LOOP THROUGH COIL

4 PULL LOOP OVER COIL

5 TIGHTEN

Figure 2-34. A coil of rope can be secured with a buntline coil for storage.

Sling Storage

Synthetic fiber slings are vulnerable to degradation from ultraviolet light so they should be stored out of sunlight and away from areas used for arc welding. Slings should be neatly hung from racks, preferably vertically, and never left in locations where vehicles or forklifts may run over them or where heavy loads may be set on them. **See Figure 2-35.** Slings should not be dragged over abrasive surfaces or sharp objects. Ideally, slings should be stored organized by size and/or type to make selection easier.

Hardware Storage

Hardware components are typically more durable than most slings and are much smaller, so they can be stored together in groups. Facilities with a lot of rigging hardware often store their hardware in one or more large cabinets or other containers commonly used for construction or maintenance equipment. Shelves and compartments in these units may be used to sort hardware by type or size. Hardware is also commonly stored on wall hooks if they can be arranged in a way that avoids tangling.

SLING STORAGE

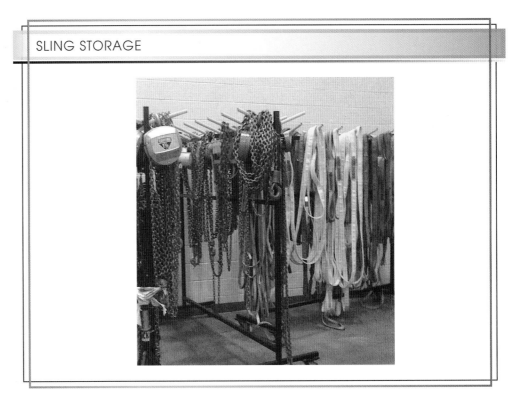

Figure 2-35. Equipment should be stored in a neat and organized manner that keeps the items clean and dry.

COMMON APPLICATIONS—
TRANSMISSION TOWER ASSEMBLY WITH HELICOPTER CRANES

Helicopter cranes may be used during the installation of high-voltage transmission towers. Helicopter cranes are capable of lifting loads of up to 25,000 lb. Transmission towers may be preassembled in sections on the ground and then erected with the use of a helicopter crane.

Southern California Edison

To assemble a transmission tower with a helicopter crane, apply the following procedure:

1. Prepare the site for the transmission tower.
2. Attach a four-legged sling at the top of the base section of the tower.
3. Attach the helicopter hoist line to the sling.
4. Lift the base section of the tower onto the concrete footings and secure the base section to the footings.
5. Lift the next tower section into place and bolt it to the previous tower section.
6. Repeat until all sections are installed.

Chapter 2 Learner Resources

Lineworker Rigging Practices

FIBER ROPE KNOTS AND SPLICES 3

OBJECTIVES
- Identify the types of fibers used in making fiber rope.
- List and describe the ways fiber rope is constructed.
- Describe the factors that affect fiber rope strength.
- List the methods of binding the cut end of a fiber rope.
- Describe how to fabricate a short splice, back splice, and eye splice.
- Describe the common types of wear and damage to look for while inspecting fiber rope.
- Identify the parts of a rope involved with knot tying.
- Identify the proper knot, hitch, or bend for a given application.
- Describe lashing and stake hold-fasts.

Fiber rope is not as strong or durable as wire rope, so it has limited applications for large and heavy lifts. However, it has advantages that make it ideal for certain applications. Fiber rope is soft and flexible, which makes it possible to tie it into a wide variety of knots and hitches. Knot tying is a critical skill. Several basic knots should be mastered by all lifting personnel.

FIBER ROPE

Fiber rope is inexpensive, flexible, and will not gouge or mar load surfaces. However, while some types of fiber rope and fiber rope slings are rated for lifting applications, they are not commonly used in industrial or construction settings. The primary disadvantage is that fiber rope is not as durable as wire rope, and it can be easily cut or frayed by sharp edges or abrasive surfaces. It is also not as strong as wire rope of similar size.

Fiber ropes do have useful applications in the lifting industry however. They are commonly used for relatively small, simple lifts. Also, since they can be easily knotted, fiber ropes are often used to secure loads to one another or to objects. For large and safety-critical lifts, fiber ropes are generally only appropriate for use as tag lines. **See Figure 3-1.** However, this application still requires knowledge of fiber rope characteristics and skill in tying effective knots. Therefore, it is important for lifting personnel to study and work with fiber ropes regularly.

FIBER ROPE AS TAG LINE

Figure 3-1. Fiber ropes are most commonly used as tag lines.

Fibers

Fiber rope is broadly classified by the material, either synthetic or natural, used to construct it. A variety of synthetic materials may be used for lifting ropes, but the most common materials are nylon, polyester, and polypropylene. **See Figure 3-2.** Some ropes are made from a single material, and some use a combination of materials, such as nylon and polyester. Synthetic ropes are nearly universal in the lifting industry because of their consistent characteristics and special properties, such as chemical resistance. Also, synthetic fibers are stronger than natural fibers. However, synthetic fibers tend to be more slippery than natural fibers, which can affect the holding power of some knots, hitches, and splices.

Natural fibers are processed from plants. Natural fibers used in manufacturing rope include manila, hemp, cotton, and sisal. Manila fiber may be used for lifting ropes, and sisal is sometimes used as a core material in wire rope. The growth and health of a living plant varies, and therefore so does the quality of the harvested fibers. This quality affects the properties of the finished rope. The resulting rope is classified by the quality of fiber used. The four common manila rope classifications include yacht rope, number 1, number 2, and hardware. Only yacht and number 1 may be used for lifting applications. A disadvantage of natural fibers is that they are vulnerable to mildew, rot, and decay.

FIBER ROPE

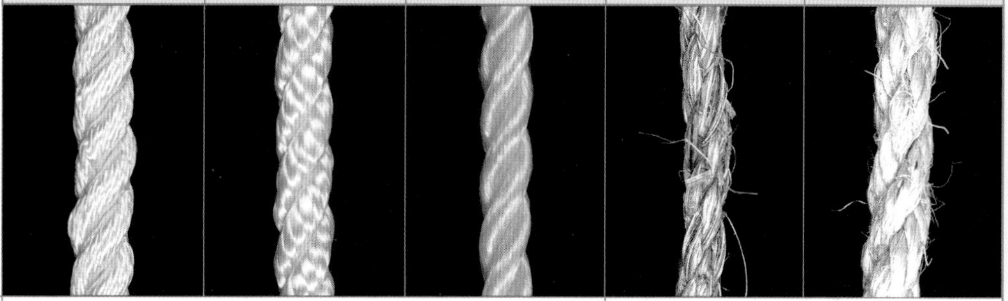

	Nylon	Polyester	Polypropylene	Manila	Sisal
Type		Synthetic		Natural	
Rot resistance	Excellent	Excellent	Excellent	Poor	Poor
Mildew resistance	Excellent	Excellent	Excellent	Poor	Poor
Oil and gas resistance	Excellent	Excellent	Excellent	Fair	Fair
Acid resistance	Good	Excellent	Excellent	Poor	Poor
Handling	Excellent	Excellent	Good	Fair	Poor
Durability	Excellent	Excellent	Good	Good	Poor
Abrasion resistance	Excellent	Excellent	Fair	Good	Fair
Dynamic load resistance	Excellent	Good	Fair	Fair	Poor
Sunlight resistance	Fair	Excellent	Poor	Excellent	Excellent
Buoyancy	Sinks	Sinks	Floats	Sinks	Sinks

Figure 3-2. Fiber rope includes synthetic and natural fiber types, each with different characteristics.

Fiber Rope Construction

Twisted fiber rope, similar to wire rope, is constructed by twisting fibers into yarn, yarn into strands, and strands into rope. **See Figure 3-3.** A *yarn* is a continuous line of fibers twisted together. A yarn is made by twisting the fibers clockwise. Several yarns are then twisted together counterclockwise to create a strand. Three or more strands are then twisted clockwise to create the rope. Reversing the twist of each step prevents the rope from unwinding. Twisted fiber rope may or may not have a core.

TWISTED FIBER ROPE

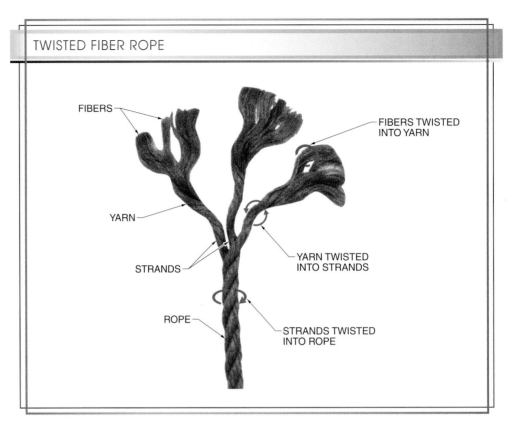

Figure 3-3. Twisted fiber rope is composed of twisted strands, each of which is composed of twisted yarns, each of which is composed of twisted fibers.

Synthetic fiber rope can also be constructed by braiding or plaiting, instead of twisting, the strands. **See Figure 3-4.** *Braiding* is the weaving of three or more untwisted strands into a rope. Various braiding patterns can be used to create hollow ropes, solid ropes, and wide, ribbonlike bands.

Single-braid rope consists of an even number of strands woven into a tube shape. This braid may be used to cover a core material, or the interior may be left void. Double-braid rope consists of completely separate inner and outer braids, which may be of different materials. Often, the inner material has greater strength while the outer material has better abrasion resistance. In both single and double braids, half of the strands spiral clockwise while the other half of the strands spiral counterclockwise. Alternatively, solid-braid rope uses strands that all spiral in the same direction but alternate between the inner and outer layers.

Plaiting is the weaving of four pairs of alternately twisted strands into a rope. Two right-hand pairs of strands are twisted counterclockwise and two left-hand pairs are twisted clockwise. This results in inherently nonrotating rope.

Fiber Rope Length and Diameter

Fiber rope can stretch considerably under tension, which lengthens it and reduces its diameter. Also, it is relatively soft and flexible. These qualities complicate the precise measurement of a rope for identification and inspection purposes. When measuring a rope, it should be laid out straight on a table or clean floor.

Fiber rope diameter specifications are given in nominal sizes, or close approximations of the average size. A *nominal value* is a specified value that may vary slightly from the actual value. Most likely, the actual diameter of a used rope will be slightly smaller than the nominal diameter, due to settling of the fibers and minor, permanent elongation. New ropes are typically closer to the nominal value.

BRAIDED AND PLAITED ROPE

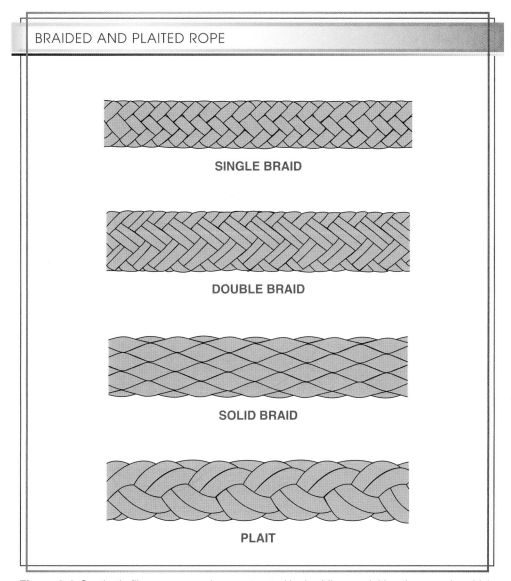

SINGLE BRAID

DOUBLE BRAID

SOLID BRAID

PLAIT

Figure 3-4. Synthetic fiber ropes can be constructed by braiding or plaiting the strands, which may or may not be twisted.

Rope length is measured with lays. **See Figure 3-5.** A *lay* is a length of rope in which a strand makes one complete spiral wrap or braiding/plaiting pattern. (This meaning for the word "lay" is in addition to the meaning indicating the direction of twist.) This unit of measure is consistent and relatively easy to identify.

For twisted rope, a strand can be traced around the rope until it is on the same side of the rope as the starting point. In braided or plaited rope, it is difficult to follow a single strand through a pattern. Therefore, rope manufacturers often include one or two differently colored strands that appear on the surface of the weave at regular intervals. Each interval is a lay that can be used to determine length.

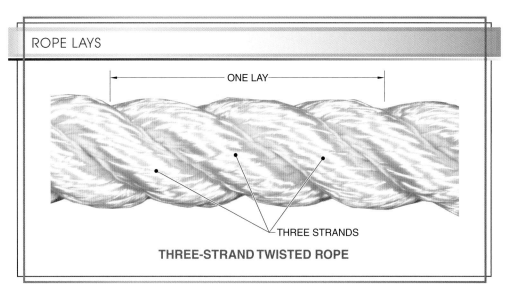

ROPE LAYS

ONE LAY

THREE STRANDS

THREE-STRAND TWISTED ROPE

Figure 3-5. A lay is a unit of length that can be used to determine how much a rope has stretched due to age or use.

Fiber Rope Strength

There is a significant difference between the strengths of different ropes, depending on size, fiber material, and construction type. **See Figure 3-6.** Generally, synthetic fiber ropes tend to be stronger than natural fiber ropes. This is partly because the synthetic materials are stronger and partly because synthetic fibers are continuous throughout the length of a rope while natural fibers are short lengths spun together.

BREAKING STRENGTHS OF SELECTED FIBER ROPES

Diameter†	Manila*	Polypropylene*	Polyester*			Nylon*		
	3-Strand Twisted	3-Strand Twisted	3-Strand Twisted	Double-Braided	8-Strand Plaited	3-Strand Twisted	Double-Braided	8-Strand Plaited
⅝	405	720	800	—	—	950	—	—
¾	540	1130	1200	1900	2000	1500	1785	1500
1	900	1710	2000	2935	3100	2600	2835	2500
1⅛	1215	2440	2800	4245	4500	3300	4095	3700
1¼	1575	3160	3800	5730	6000	4800	5355	5000
1½	2385	3780	5000	7500	7700	5800	7245	6400
1¾	3105	4600	6500	9450	9700	7600	9450	8000
2	3960	5600	8000	11,660	12,100	9800	12,600	11,000

* in lb
† in in.

Figure 3-6. Synthetic fiber ropes are generally stronger than natural fiber ropes, though the strength varies with construction type.

As with wire rope, fiber rope strength is provided as a breaking strength. This does not account for a safety factor, since the appropriate safety factor may vary depending on the application. In most cases, though, a safety factor of 5 is used. This makes the rated loads one-fifth of the breaking strengths. Other factors related to the use and conditions of a rope may affect the final rated load.

When used with sheaves (pulleys), fiber rope strength should be derated for the weakening effect of bending. Just as for wire rope, the bend ratio can be calculated from the sheave and rope diameters. Then the bending efficiency for the rope can be researched in the manufacturer's literature.

Knot Efficiency. Unlike wire rope, fiber rope can be tied into knots. As with bending, the extreme turns and pinches resulting from knots can reduce the strength of a rope significantly. *Knot efficiency* is the ratio of the strength of a knotted rope to its nominal strength rating. **See Figure 3-7.** That is, a rope with a knot with a 60% efficiency retains only 60% of the strength of the same rope without a knot.

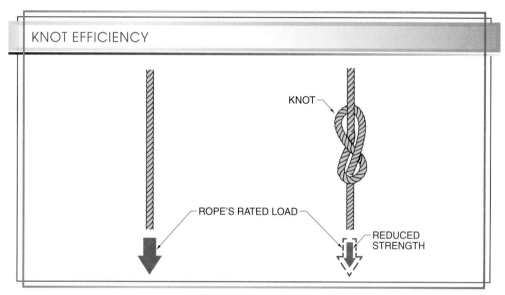

KNOT EFFICIENCY

KNOT

ROPE'S RATED LOAD

REDUCED STRENGTH

Figure 3-7. A knot reduces the strength of a rope. Knot efficiency is the percentage of rated load that the rope retains after being knotted.

Knot efficiency depends on many factors, however, such as fiber type, rope type, rope size, rope condition, manner of loading, and knot dressing. *Dressing* is the adjustment of a knot as it is tightened for a clean look and tight arrangement. Therefore, efficiency values can only be approximations and should be used cautiously. These values are most useful as relative values for comparing knots with each other. For example, a knot with a 60% efficiency rating probably varies between 50% and 65% efficiency in actual use, but it is still likely to be more efficient than a knot with a 50% efficiency rating.

Environmental Conditions. Environmental conditions are major factors in fiber rope strength, even more than for other types of lifting ropes and slings. Natural fiber ropes are particularly vulnerable to environmental damage, though synthetic ropes are also susceptible, especially from extreme temperatures.

Natural fibers can absorb moisture, which leads to decay or rot. Therefore, most rope manufacturers treat natural fiber rope with waterproofing. However, enough moisture may still be absorbed to significantly weaken a natural fiber rope, particularly when frozen. Natural fiber rope must be completely thawed before use. Most synthetic fibers do not absorb moisture, but the rope may still become brittle and weakened if coated with ice.

Fiber ropes are typically rated for use in the temperature range of –20°F to 180°F (–29°C to 82°C). High temperatures dry out natural fibers, making them brittle and easily breakable. Synthetic fibers tend to soften in high temperatures, allowing the rope to stretch and lose strength. Very low temperatures tend to make either type of fiber brittle.

Natural fiber ropes cannot be used in chemical environments because natural fibers deteriorate quickly. Synthetic fibers withstand many chemical environments, making synthetic fiber ropes particularly well-suited for applications in these environments. However, resistances vary, so the manufacturer's literature should always be consulted before using a rope in a chemical environment.

Whipping

Similar to wire rope, fiber ropes should be bound at the ends before cutting. As fiber ropes are more likely to unravel than wire ropes, bindings are critical. A number of methods can be used to bind ends. A traditional method is called whipping, which is similar to seizing wire ropes. *Whipping* is the wrapping of twine that binds the end of a fiber rope near where it is cut. **See Figure 3-8.**

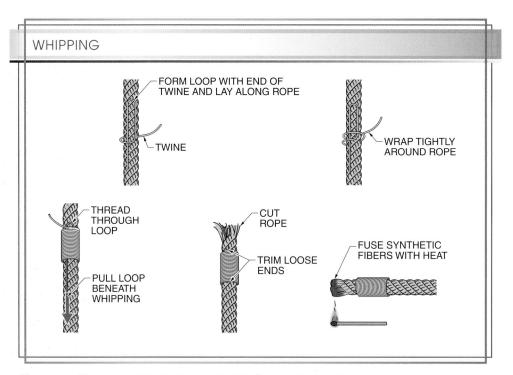

WHIPPING

FORM LOOP WITH END OF TWINE AND LAY ALONG ROPE

TWINE

WRAP TIGHTLY AROUND ROPE

THREAD THROUGH LOOP

PULL LOOP BENEATH WHIPPING

CUT ROPE

TRIM LOOSE ENDS

FUSE SYNTHETIC FIBERS WITH HEAT

Figure 3-8. Fiber rope whipping is completed before cutting in order to bind the strands together.

The whipping material is flexible twine or thin strands of synthetic fibers. First, the twine is formed into an elongated loop, which is laid along the rope to be whipped. Then the twine is wrapped tightly around the rope and over the loop while gradually worked toward the rope end. The turns are laid hard against each other without overlapping. When the whipping is of sufficient length, the loose end of the twine is threaded through the remaining part of the loop. The loop is then drawn beneath the whipping by pulling the other end of the twine. Both loose ends of the twine are trimmed close to the turns, and the rope is then cut. After cutting, the ends of some synthetic fiber ropes are also fused with heat.

A number of alternative binding methods can also be used. **See Figure 3-9.** A common method is to wrap an end in self-adhesive plastic tape, usually electrical tape. Similarly, a short length of heat-shrink tubing can be threaded onto the rope and shrunk over the end. Another method is to apply a special liquid coating to the last inch or two of a rope end. The coating forms a rubbery, flexible coating on the end when dry. Sometimes the end of a rope is simply fused with heat, which is effective if done thoroughly. However, it can be difficult to ensure that all the fibers around the edge are completely fused. Finally, the end of a rope can be rewoven into a back splice, which forms a smooth end with a thicker diameter for the last few inches.

WHIPPING AND OTHER END BINDINGS

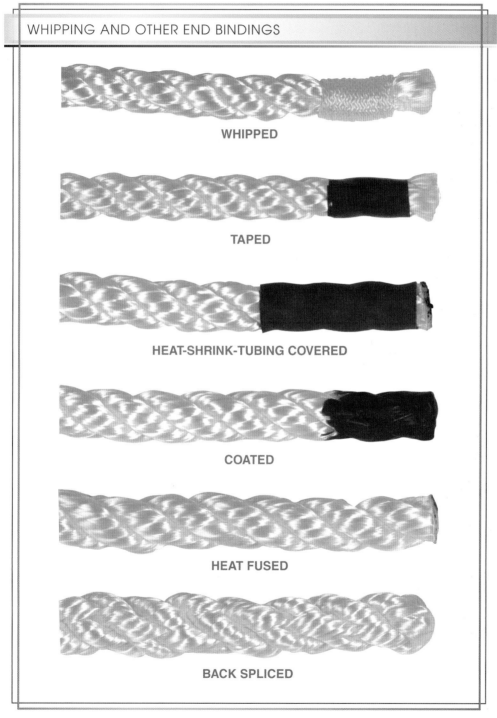

WHIPPED

TAPED

HEAT-SHRINK-TUBING COVERED

COATED

HEAT FUSED

BACK SPLICED

Figure 3-9. Besides whipping, there are other common methods of binding the end of a cut rope to prevent unraveling.

Fiber Rope Splicing

A *splice* is the unlaying and then reweaving together of two portions of rope in order to form a permanent connection. *Unlaying* is the untwisting of a rope's strands. Splices are commonly used to join two ends of rope. If the ends are from separate ropes, the ropes must be of similar strength and thickness.

Splices retain far more of a rope's original strength than knots do. Theoretically, a perfect splice has an efficiency nearing 100%. In actual practice, most splices are not perfect but still achieve efficiencies in the range of 85% to 95%. Therefore, if a semipermanent splice is an option for a certain application, it is preferable to a knot.

Ropes are spliced by unlaying a portion of both rope ends and then tucking the loose strands of each end into the twisted strands of the other. **See Figure 3-10.** The minimum number of tucks may vary depending on the rope and splice type, but the minimum is typically four for natural fiber ropes or five for synthetic fiber ropes. (Synthetic ropes have a higher minimum because they tend to be more slippery.)

SPLICE TUCKS

Figure 3-10. Ropes are spliced by tucking the loose strands of one rope, or portion of rope, into the twisted strands of another rope, or portion of rope.

There are different methods for determining the length of rope that should be unlaid. One rule of thumb is three times the rope diameter for each tuck. For example, a ½″ rope requiring five tucks should be unlaid about 7½″. Alternatively, another rule of thumb is one lay per tuck, plus a little extra. For example, for five tucks, the end is unlaid for a length equal to about six lays. However, it is better to have too much rope than not enough. If the unlay is too short for the required number of tucks, the splice must be completely rewoven. If it is too long, however, the extra length can either be cut off after the necessary tucks are completed or woven as extra tucks.

Only three-stranded twisted rope can be practically spliced by most field personnel. The most common types of splices are short splices, eye splices, and back splices. The result of splicing is a thickened portion of rope where six strands are woven together.

After a minimum number of tucks, all three strands are cut as needed, fused with heat, and tucked into the standing part. This abruptly terminates the thick weave.

Alternatively, all three splice types allow the option of tapering the splice. **See Figure 3-11.** To follow this procedure, the strands are successively trimmed as they are woven back into the rope. After the minimum number of tucks, the first strand is cut off, fused, and tucked. The remaining strands are tucked one more time and then another strand is cut, fused, and tucked. Finally, the last strand is tucked one more time and then cut, fused, and tucked. The result is a splice that gradually changes from a thick, six-strand weave down to the original, thinner, three-strand lay.

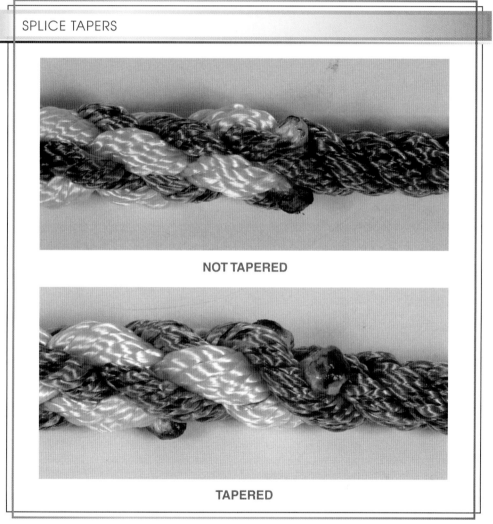

SPLICE TAPERS

NOT TAPERED

TAPERED

Figure 3-11. A thick section of spliced rope can be terminated abruptly or tapered down to the original rope thickness.

Short Splices. A short splice is used to join two ends of rope. **See Figure 3-12.** A short splice is often used to permanently join two ends of a single piece of rope together, forming a roundsling. One application of this is rigging a gin wheel, which is a simple pulley for manually lifting small loads. Since the rope is a continuous loop, it is easy to either hoist or lower a load by pulling on one side or the other.

For additional information, visit qr.njatcdb.org Item #1650

SHORT SPLICES

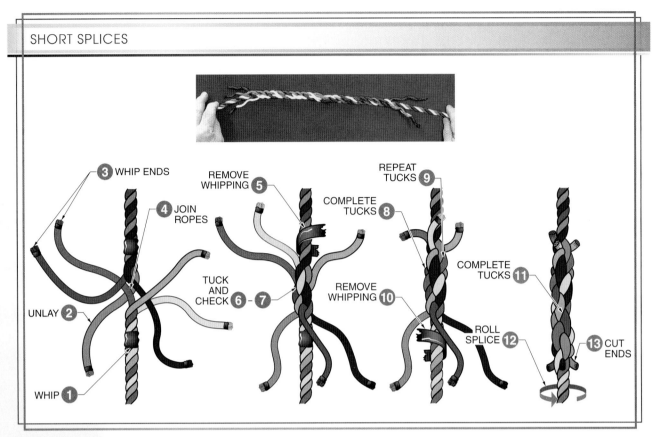

Figure 3-12. A short splice is a method for joining two rope ends.

When completed, a short splice is about twice as thick as the rest of the rope. A disadvantage is that this thickness makes it vulnerable to riding out of a sheave groove or binding up within a block. A short splice is formed using the following procedure:

1. Place temporary whipping (usually tape) on each rope at a distance from the end equal to the unlaying length.
2. Unlay rope on each end.
3. Place temporary whipping on the strand ends.
4. Push the two rope ends into each other, alternating the strands of one end with the strands of the other.
5. Remove the temporary whipping from one rope.
6. Tuck each strand from the whipped rope into the V rope. Each strand from the whipped rope should cross over one strand and then tuck under the next in the unwhipped rope. Complete one tuck for each strand of that rope.
7. Ensure all tucks are snug and placed correctly. Each whipped rope strand should alternate with unwhipped rope strands.
8. Complete another set of tucks for each strand. Check and snug the splice.
9. Repeat for the required number of tucks.
10. Remove the temporary whipping from the other rope.
11. Complete the required number of tucks on the other side of the splice, checking and snugging after each set.

12. Roll the splice under pressure to ensure that each strand settles into position.

13. Cut all the strands to the same length, or taper by alternating cutting one strand and tucking the remaining strands. Fuse each cut with heat or some other permanent means and complete the final tuck.

Back Splices. A back splice, also known as a crown, finishes a rope end by braiding its loose strands back on itself. **See Figure 3-13.** This is an alternative to whipping when an enlarged rope end is desired or, at least, not objectionable. This is commonly done to tag lines on a working end in order to mark the end of the rope. If the rope is running through the hands of a tag line holder, that person will feel the end approaching without having to look down. The back splice becomes tighter with time and use. A back splice is formed using the following procedure:

For additional information, visit qr.njatcdb.org Item #1651

1. Place temporary whipping (usually tape) on the rope at a distance from the end equal to the unlaying length.

2. Unlay rope on the end.

3. Place temporary whipping on the strand ends.

4. Loop strand 1 and hold its end to maintain the loop.

5. Pass strand 2 through the loop of strand 1 and behind strand 3.

6. Lay strand 3 over strand 1 and through the loop formed by strand 2.

7. Tighten the knot by snugging all three strands. This forms a crown knot.

8. Tuck each strand into the rope. Each strand should cross over one strand and then tuck under the next. Complete one tuck for each strand of the rope.

9. Ensure all tucks are placed correctly. Each working part strand should alternate with standing part strands.

10. Complete the required number of tucks for each strand, checking and snugging after each set.

11. Roll the splice under pressure to ensure that each strand settles into position.

12. Cut all the strands to the same length, or taper by alternating cutting one strand and tucking the remaining strands. Fuse each cut with heat or some other permanent means and complete the final tuck.

Eye Splices. Splices are also used to join portions of the same rope. An eye splice forms a loop at the end of a rope by splicing the rope end into the rope body. Eye splices typically contain a thimble for protecting the rope. **See Figure 3-14.** The thimble is inserted after the splice is completed and may be held in place by whipping at the base of the eye. An eye splice is formed using the following procedure:

For additional information, visit qr.njatcdb.org Item #1652

1. Place temporary whipping (usually tape) on the rope at a distance from the end equal to the unlaying length.

2. Unlay rope on the end.

3. Place temporary whipping on the strand ends.

4. Form an eye of the desired size by pushing the end into the side of the rope with two strands (numbers 1 and 2) on top and one strand (number 3) underneath.

5. Tuck strand 1 (one of the top strands) under one of the strands in the standing part, perpendicular to the lay of the rope.

6. Strand 2 (the other top strand) is tucked under a standing part strand, also in a perpendicular direction. It passes over the standing part strand that strand 1 was tucked under and is tucked under the next strand.

7. Turn the assembly over.

8. Strand 3 is tucked under a standing part strand. It enters where strand 2 exits the standing part and tucks under the adjacent strand in the opposite direction.

9. Ensure all tucks are snug and placed correctly. Each working part strand should alternate with a standing part strand.

10. Complete a second tuck for each strand into the standing part. Each strand crosses over one strand and then tucks under the next.

11. Ensure all tucks are snug and placed correctly.

12. Complete the required number of tucks for each strand, checking and snugging after each set.

13. Remove all temporary whipping.

14. Optional: Insert thimble and add whipping.

15. Roll the splice under pressure to ensure that each strand settles into position.

16. Cut all the strands to the same length, or taper by alternating cutting one strand and tucking the remaining strands. Fuse each cut with heat or some other permanent means and complete the final tuck.

BACK SPLICES

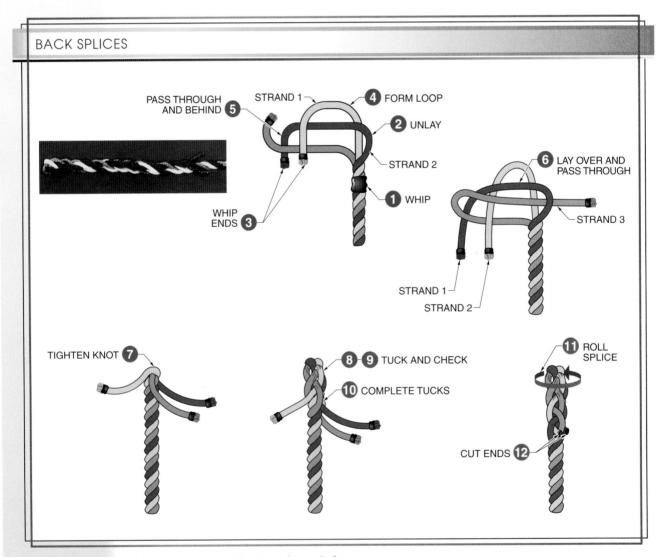

Figure 3-13. A back splice is a way to neatly dress the end of a rope.

EYE SPLICES

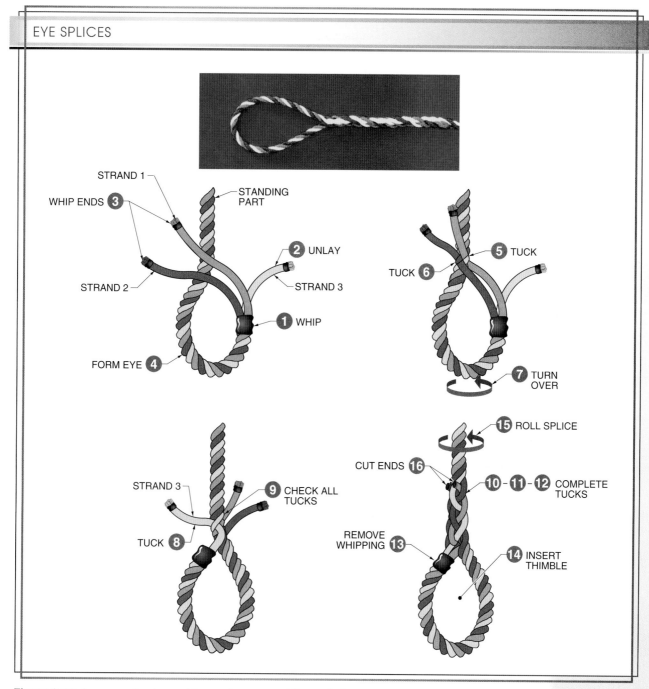

Figure 3-14. An eye splice is used to create an eye at the end of a rope.

FIBER ROPE INSPECTION

Due to its more flexible nature, fiber rope can be inspected more thoroughly than wire rope. The strands of a fiber rope can often be pulled open slightly to inspect the interior for dirt and grit, which can gradually wear fibers to the breaking point. However, the strands must be opened carefully to avoid causing distortions that reduce the strength of the rope.

Before each use, the exterior of a rope should be inspected for signs of excessive abrasion, wear, cuts, kinking, moisture damage, thermal damage, and chemical damage.

See Figure 3-15. A fiber rope may have many frayed, fuzzy surface fibers due to gradual abrasion and wear. Wear due to frequent sheave (pulley) use may cause matted or glazed fibers on the surface due to heat buildup. The level of wear considered excessive depends on the extent of the damage and the proportion of the outer strands that are worn away.

FIBER ROPE DAMAGE

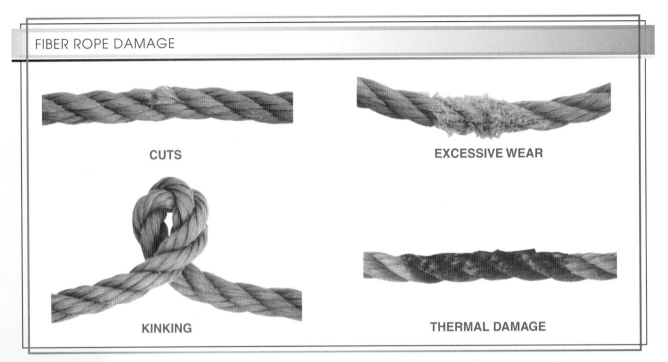

CUTS

EXCESSIVE WEAR

KINKING

THERMAL DAMAGE

Figure 3-15. Common types of damage to fiber rope include cuts, kinking, excessive wear, and thermal damage.

Cuts are localized damage caused by contact with sharp edges of loads or other equipment. A rope should be removed from service if it has cuts with more than a few broken fibers. If the damage is near an end, and the rope is otherwise acceptable, it may be possible to remove the damaged portion and use the remainder of the rope.

Kinks are often formed when loops of rope cannot untwist when pulled tight, distorting the lay of the strands at one spot. Similar distortions can also be caused by excessively twisting a rope in either direction. Minor kinks can sometimes be undone by applying an opposite twist to a rope until the rope lays flat and the strands twist uniformly. If this is not possible, the rope should be discarded. Braided and plaited ropes do not usually form kinks.

Mold or mildew may be present in a rope that is repeatedly exposed to moisture and not dried promptly. This contamination appears as black spots or areas that cannot be easily cleaned. There may also be a musty odor. Natural fiber ropes with any sign of mold or mildew should be immediately removed from service as they may be seriously weakened. Synthetic fiber ropes may not be adversely affected by mold or mildew. The rope manufacturer should be consulted on a course of action in these situations.

Thermal damage on a rope can be caused by excessive friction when it is wound around sheaves or from direct contact with flames or hot equipment. Natural fibers tend to become charred from heat, while synthetic fibers tend to melt. Small amounts of heat damage may be acceptable, but if in doubt, a rope should be discarded.

Chemicals may cause discoloration or physical damage to fibers. Discoloration may not affect rope strength at all, but physical damage to fibers can significantly weaken rope.

The degree of damage depends greatly on the chemical and type of rope fiber. Unless the exact nature of the chemical exposure is known and it is unlikely the rope is affected, then the rope should be discarded if there is any evidence of chemical exposure.

KNOT TYING

For additional information, visit qr.njatcdb.org Item #1653

Fastening loads with fiber rope normally requires tying some type of knot. A *knot* is the interlacing of a part of a rope to itself, which is then drawn tight. Knots are designed to form a semipermanent connection that can be later untied. The sharp directional changes and pinching of the rope in a knot can weaken a rope by as much as 55%.

The generic term "knot" is frequently used to describe any intentional crossing in a rope. More specific terms are used to describe the purpose for which the knot is made. A *hitch* is the binding of rope to another object, often temporarily. Without the other object, the hitch will not retain its proper form. A *bend* is a knot that is used to tie the ends of two ropes together.

There are dozens of knots, hitches, and bends that are useful for securing, binding, arranging, attaching, and lifting loads. Some, especially the most common, originated thousands of years ago and are used in many different ways, while others are newer, more specialized types. Many, especially those with long histories, have multiple names. Also, some related knots look similar but have different characteristics. Rigging personnel should be careful when studying and tying knots to use them accurately and appropriately. The improper application and execution of a knot can have disastrous consequences, and rigging personnel need to understand why one type of knot should be used instead of another.

Even though it may be easy to visualize some of the simpler knots without tying them, rigging personnel should practice forming and applying these knots as part of their training. Knot tying is a skill that involves not only knowledge of the type of knot to be used but also spatial awareness and hand-eye coordination, which only come with practice. The proper tying of certain types of knots is also a requirement for certification testing.

Knots that are unstable have a tendency to capsize and should not be used for rigging. *Capsizing* is the tendency of a knot under load to change form, often in an adverse way. Other knots can be used safely but have a tendency to jam, making them difficult to untie. Therefore, they may not be the best to use for a given application. Another consideration is whether a knot has the potential to close up around a part of workers' bodies and trap them or cause injury. This can be avoided if the proper type of knot is used.

Knots and hitches may be used when lifting and stabilizing loads.

Rope Terminology

Fiber rope work, particularly knot tying, requires some knowledge of special terminology. Names have been given to the different parts of a rope to help describe how the parts are involved in each step of a knot-tying procedure.

Most rope terminology was derived from nautical, or sailing, terms. **See Figure 3-16.** A *bight* is a loose or slack part of a rope between two ends. A *loop* is the folding or doubling of a line to create an opening through which another line may pass. A *nip* is a pressure point created when a rope crosses over itself after a turn around an object. A nip holds a knot together because its pressure, along with the friction of the rope material, keeps the rope from slipping through and loosening.

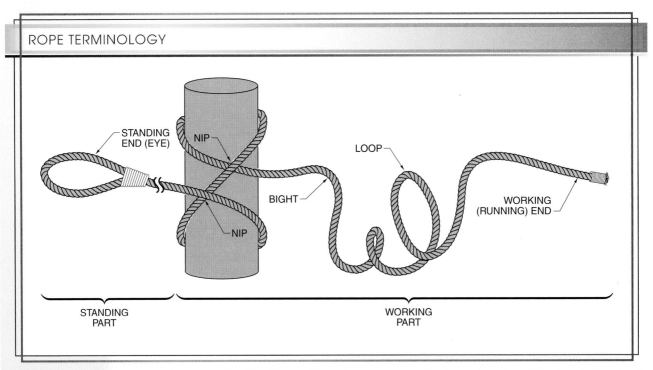

Figure 3-16. Certain terminology is used to identify specific parts of a rope, particularly when describing how to tie a knot.

The portions of rope involved are identified as working or standing. The *working part* is the portion of a rope involved in making a knot, hitch, or bend. The *working end* is the end of a working part. The working part of a rope is loose, and the end can be passed through or around other objects or the standing part of the rope. The *standing part* is the portion of a rope that is unaltered or not involved in making a knot, hitch, or bend. The *standing end* is the end of a standing part. The standing end is often attached to some other object, making it taut and unable to be worked.

Some knots are formed using the working end of a rope. This type of tying may be necessary if the rope must pass through an object, such as a ring. Other knots are formed at some point along a rope and do not require threading the rope end at all. If these knots are used with an object, the rope must be able to be passed over one end of the object. Finally, some knots can be tied either way and have two sets of tying procedures. For these knots, a distinction must usually be made between the two methods. The first type is typically called "end tying," "threading the end," or something similar, while the latter is typically called "tying on the bight," "loops tying," or something similar.

Common Rigging Knots

Rigging knots are primarily used in one of three ways: as stopper knots, to fasten two ropes together, or to form loops. A *stopper knot* is a category of knots that are used to prevent the ends of ropes from slipping through openings or other knots. Knots that fasten ropes together are typically used to bind and secure loads. Loops are typically used for attaching ropes to other objects. There are hundreds of different knots and variations, but only a few are needed for most applications. Rigging personnel should be well-versed in the common rigging knots.

Overhand Knots. The overhand knot is a fundamental knot and forms the basis for many other knots. **See Figure 3-17.** This knot is very secure to the point that it is difficult to untie after it has been tightened under tension. Its knot efficiency is about 50%. If the overhand knot is slid over an object as tension is applied to the ends, a marline (also spelled "marlin") hitch is formed. Marline hitches, or half hitches, can be used in conjunction with a spar hitch or rolling hitch to help control a load when hoisting a long object vertically. They can also be used to bind rolled up materials, such as tarps, carpeting, or linoleum. An overhand knot is tied by applying the following procedure:

1. Form a loop with the working part over the standing part.
2. Pass the working end underneath the loop and back up through it.
3. Tighten by pulling on the working end.

For additional information, visit qr.njatcdb.org
Item #1654

OVERHAND KNOTS

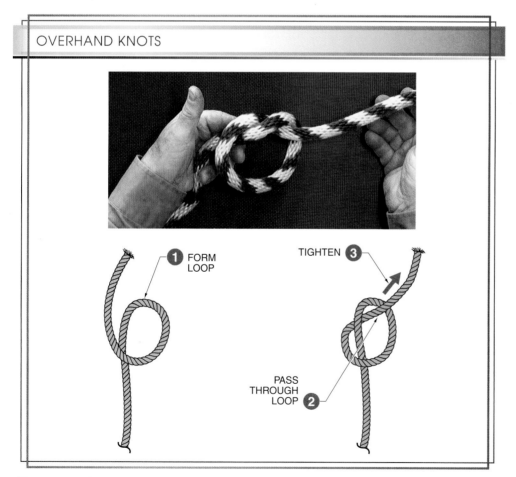

Figure 3-17. An overhand knot is a simple, effective knot and is difficult to untie.

Square Knots. A square knot, also known as a reef knot, is formed by tying together two ends of the same rope, such as when a rope is wrapped around an object and the two ends meet and are knotted together. **See Figure 3-18.** While common, a square knot is not very secure and should never be used to fasten two different ropes together or for any critical application. It may, however, be used for noncritical fastening, such as tying down a tarp. This knot must be mastered in order to pass certain rigging exams, but its limited utility and likelihood of tying incorrectly make it a knot to be avoided for everyday rigging. It is about 45% efficient.

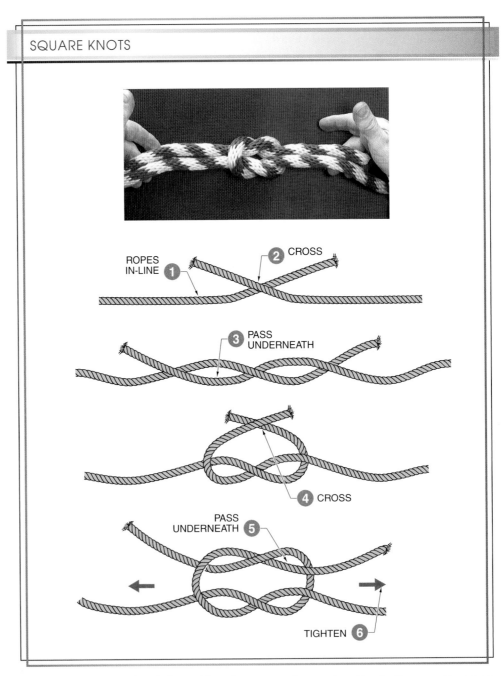

Figure 3-18. A square knot is formed from two successive overhand knots. Care must be taken to tie the knot properly to avoid inferior variations.

A square knot is formed by a combination of two overhand knots, though special attention should be paid to the way the ends are crossed. When done improperly, it forms a granny knot, which is inferior in holding power and can release suddenly and unexpectedly. When done properly, the knot will lay relatively flat and be symmetrical. Also, each working end will lay next to its standing part. If tied improperly, the ends come out at right angles to the standing parts. A square knot is tied by applying the following procedure:

1. Arrange the two rope ends side by side but pointing in opposite directions.

2. Cross one end over the other.

3. Pass the end of the top rope underneath and around the bottom rope. This forms a loose overhand knot.

4. Cross one end over the other. The top rope must be the same rope that was on top in step 2.

5. Pass the end of the top rope underneath and around the bottom rope.

6. Tighten by pulling on the working ends.

Bowline Knots. A bowline knot, which is pronounced "BO lin," is one of the most important and versatile knots. It forms a loop that is fixed in size at the end of a rope. **See Figure 3-19.** This knot has an efficiency of about 75%. As more strain is placed on the rope, the knot becomes tighter, but the loop does not constrict. When not under strain, the knot is easily released. In fact, the primary disadvantage of the bowline knot is its tendency to work loose when not under load. This can be prevented by taping the end of the rope to the side of the loop.

The bowline knot has many different uses. It is often used to attach a tag line to a load and then to an empty crane hook for a return trip to the ground. It can also be used as a slipknot by forming the eye around the standing part of the rope. In this form, the knot can be used for binding up a load, such as a bundle of pipe or lumber.

For additional information, visit qr.njatcdb.org Item #1656

BOWLINE KNOTS

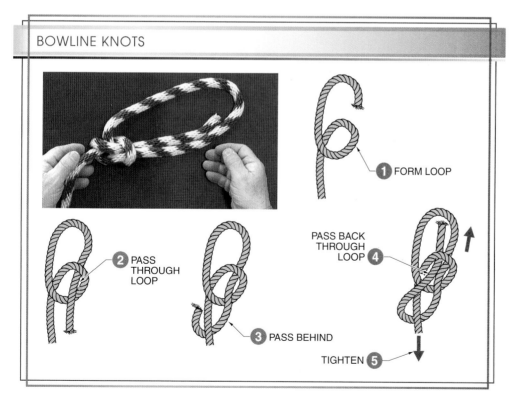

① FORM LOOP

② PASS THROUGH LOOP

③ PASS BEHIND

PASS BACK THROUGH LOOP **④**

TIGHTEN **⑤**

Figure 3-19. A bowline knot forms a fixed loop at the end of a rope and is one of the most useful knots.

BOWLINE ON A BIGHT

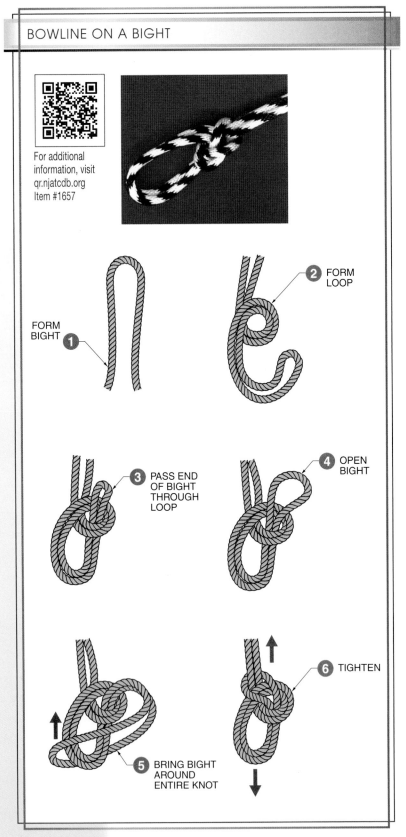

For additional information, visit qr.njatcdb.org Item #1657

Figure 3-20. A bowline on a bight is a bowline knot tied in the middle of a line.

Note that the bowline knot can be tied with the working end on either the inside or the outside of the loop, though the knot is stronger with the working end on the inside. A bowline knot is tied by applying the following procedure:

1. Form a loop with the working part over the standing part. Allow for enough working part to form the size of the loop desired at the end of the rope.
2. Thread the working end beneath and through the loop.
3. Pass the working end under the standing part.
4. Pass the working end over the standing part and back through the loop.
5. Tighten by pulling the standing part and the working end.

A *one-handed bowline* is a bowline knot that is tied with only one hand. A one-handed bowline is tied by applying the following procedure:

1. Hold the working end and hook the standing part with the thumb to form a loop around the hand.
2. Pass the working end around the standing part.
3. Pull the working end back through the loop while removing the hand from the loop.
4. Tighten by pulling the standing part and the working end.

A *bowline on a bight* is a bowline knot tied in the middle of a line to tie the line around an object. **See Figure 3-20.** A bowline on a bight is tied by applying the following procedure:

1. Form a bight in the middle of a piece of rope by folding a section of the rope over itself.
2. Form a loop.
3. Pass the end of the bight through the loop.
4. Open the bight.
5. Bring the bight around the entire knot until it encircles both standing parts.
6. Tighten by pulling on the loop and standing parts.

A *running bowline* is a bowline-knot-and-noose combination that does not bind, slides easily, and can be undone easily. **See Figure 3-21.** A running bowline is tied by applying the following procedure:

1. Pass the working end around the object.
2. Form a loop on the working end.
3. Pass the working end around the standing part and through the loop.
4. Pass the working end around itself and back through the loop.
5. Pull on the standing part to tighten against object.

For additional information, visit qr.njatcdb.org
Item #1658

RUNNING BOWLINE

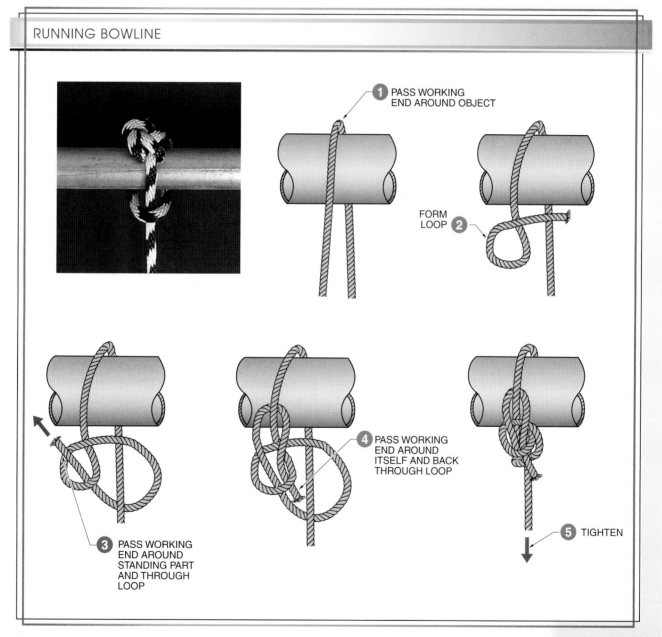

Figure 3-21. A running bowline is a bowline knot that is a type of noose that does not bind, slides easily, and can be undone easily.

A *three-ring bowline* is a bowline knot that is formed quickly with three loops of rope. **See Figure 3-22.** A three-ring bowline is tied by applying the following procedure:

1. Hold one hand up, with the rope crossing the width of the palm.
2. Wrap the rope around the hand to form three loops.
3. Take the right loop and pass it under the other two loops.
4. Pass the center loop under the right loop.
5. Pass the new center loop under the left loop.
6. Tighten by pulling on the loop and standing parts.

THREE-RING BOWLINE

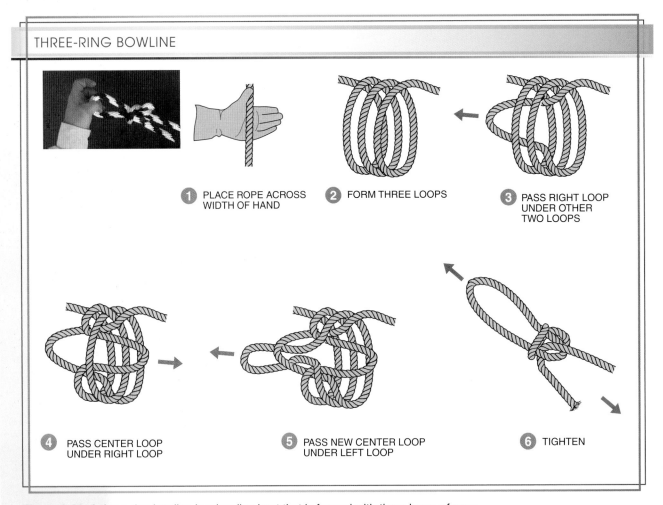

Figure 3-22. A three-ring bowline is a bowline knot that is formed with three loops of rope.

A *triple bowline* is a bowline knot that has three loops and is typically used to help with lifting injured personnel. A triple bowline is tied by applying the following procedure:

1. Double a length of rope.
2. Form a counterclockwise loop with the doubled line.
3. Pass the end of the bight through the loop so that it comes out over the top of the loop.
4. Pass the end of the bight behind the standing part and back through the loop.
5. Tighten by pulling after adjusting the size of the loops.

A *double bowline* is a bowline knot that has an extra loop in the knot. A double bowline is stronger and more secure than a standard bowline. **See Figure 3-23.** A double bowline is tied by applying the following procedure:

1. Form two overhand loops.
2. Pass the working end through the two loops from behind.
3. Pass the working end behind the standing part.
4. Pass the working end back through the loops.
5. Tighten by pulling on the loops and standing part.

For additional information, visit qr.njatcdb.org Item #1660

DOUBLE BOWLINE

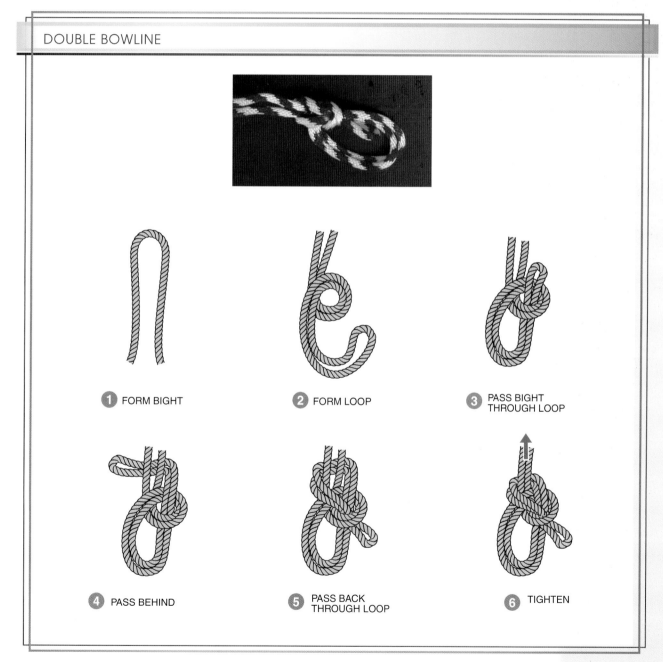

Figure 3-23. A double bowline is a bowline knot with an extra loop in the knot.

Grunt's Knots. A grunt's knot is used for sending light objects or equipment, such as a hard hat, up a utility pole via a handline. **See Figure 3-24.** The grunt's knot can be untied from the ground by pulling on the rope. A grunt's knot is tied by applying the following procedure:

1. Form a bight in a length of rope.
2. Pass the bight of rope through a light object.
3. Form a half hitch on the standing part of the rope.

GRUNT'S KNOTS

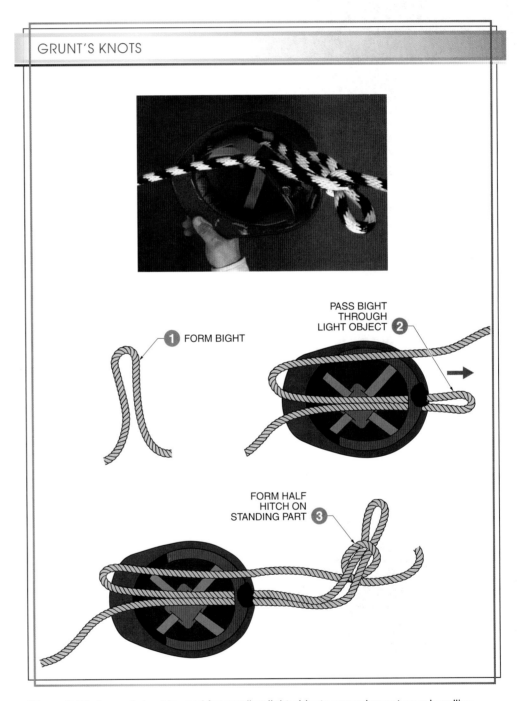

Figure 3-24. A grunt's knot is used for sending light objects or equipment up a handline.

Shoestring Knots. A shoestring knot is used on a smaller diameter rope for stabilizing a crossarm being sent up a utility pole via a handline. **See Figure 3-25.** A shoestring knot is tied by applying the following procedure:

1. Wrap a smaller rope around the crossarm twice.
2. Tie a half knot by crossing one end over the other and passing the end of the top rope underneath and around the bottom rope.
3. Form a loop with one end.
4. Wrap the other end around the loop.
5. Form a second loop by pulling part of the rope through the wrap.
6. Tighten by pulling both loops.

For additional information, visit qr.njatcdb.org Item #1662

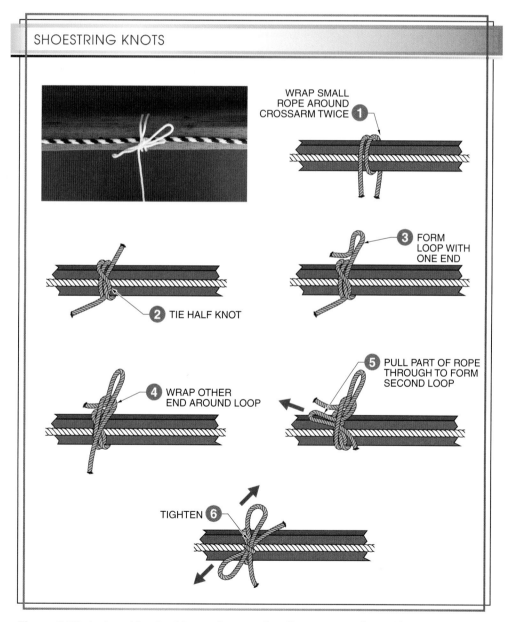

Figure 3-25. A shoestring knot is used on smaller diameter rope for stabilizing a crossarm being sent up a handline.

Lineman's Loops. A lineman's loop is a convenient way to add a fixed loop to the middle of a rope, that is, when neither end of a rope is free. **See Figure 3-26.** It is commonly used in climbing applications, where a carabiner can be attached to the loop. In this application, it is commonly called a butterfly loop or butterfly knot. For rigging, this knot can be handy as a hand-hold or tie-off point in the standing part of a rope for binding or adjusting a load. The efficiency of this knot is about 70%. It is a secure knot that can be released when not under load. The advantage of the lineman's loop is that it can be loaded from the standing part, the working part, or the loop, and it will not capsize or collapse.

The simplest method of tying this knot is to start by winding the loops around a hand. It is possible to form this knot in other ways, but a hand helps to hold the loops in place until the knot can be tightened. A lineman's loop is tied by applying the following procedure:

1. Hold one hand up, with the rope crossing the width of the palm.
2. Grasp one end of the rope and pass it behind the hand.
3. Loop around the finger tips and then back behind the hand.
4. Bring the rope back down and across the width of the palm.
5. Pull the fingertip loop down over the two sections of rope across the palm.
6. Push the loop up behind the palm sections and slide the knot off the hand.
7. Tighten the knot while maintaining the loop.

LINEMAN'S LOOPS

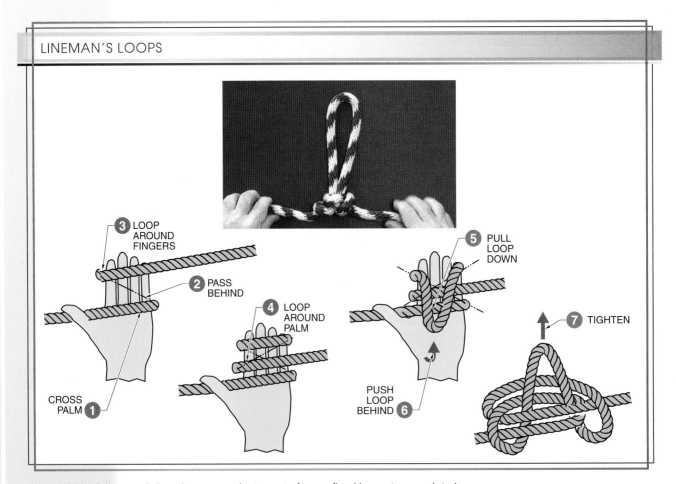

Figure 3-26. A lineman's loop is a convenient way to form a fixed loop at any point along a rope.

Figure Eight Knots. The figure eight knot is commonly used as a stopper knot. **See Figure 3-27.** This knot is more reliable than an overhand knot but is still vulnerable to unintentionally releasing when under pressure. It is also known as a Flemish knot and has an efficiency of about 80%. A figure eight knot is tied by applying the following procedure:

1. Form a loop with the working part over the standing part.
2. Pass the working part under the standing part.
3. Bring the working end up and pass it down through the loop.
4. Tighten by pulling on the working end.

For additional information, visit qr.njatcdb.org Item #1664

FIGURE EIGHT KNOTS

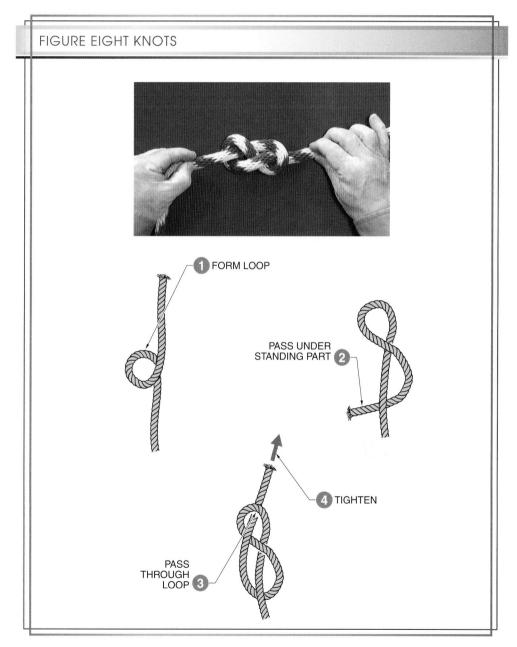

Figure 3-27. A figure eight knot is an effective stopper knot.

For additional
information, visit
qr.njatcdb.org
Item #1665

Figure Eight Loops. A figure eight loop, also known as a Flemish loop, is a strong knot used to form a loop at either the end or the middle of a rope. **See Figure 3-28.** It is secure but tends to be difficult to untie after being loaded. It is tied in exactly the same way as a figure eight knot, except the knot is tied with a bight instead of a working end. A figure eight loop on the bight is tied by applying the following procedure:

1. Fold the rope and hold the two lengths together while tying.
2. Form a loop with the working part (folded bight) over the standing part.
3. Pass the folded bight under the standing part.
4. Bring the folded bight up and pass it down through the loop.
5. Tighten by pulling on the loop.

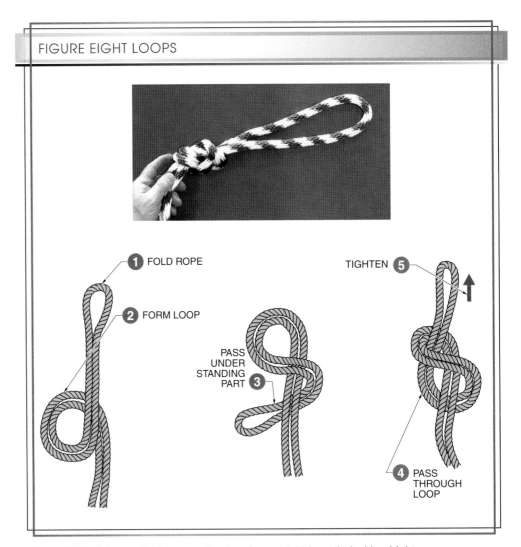

Figure 3-28. A figure eight loop is simply a figure eight knot tied with a bight.

Common Rigging Bends

Tying two ropes together can be accomplished by using a bend. A bend is typically used when a smaller rope is used to hoist or pull a larger rope. Electricians frequently use a similar technique by blowing a pulling string through a conduit and then gradually pulling in larger lines until the final pulling rope is in the conduit. While there are many different types of bends, the most practical ones for rigging purposes are the sheet bend and the double sheet bend.

Sheet Bends. A sheet bend is related to the bowline knot but is used as a bend. **See Figure 3-29.** The sheet bend is used for joining ropes of slightly different diameters. It is a secure knot under load but has a tendency to easily work loose when not under load. Its efficiency is about 55%. When tied correctly, both rope ends should lay on the same side of the knot. Otherwise, the resulting knot is an inferior variation that is not as secure. A sheet bend is tied by applying the following procedure:

1. Fold the thicker rope into a flat bight.
2. Pass the end of the thinner rope up through the bight's loop.
3. Pass the thinner rope around behind the folded working end and standing part of the thicker rope.
4. Bring the end of the thinner rope up to the front and then pass it underneath itself.
5. Tighten by pulling on the thinner rope.

For additional information, visit qr.njatcdb.org Item #1666

SHEET BENDS

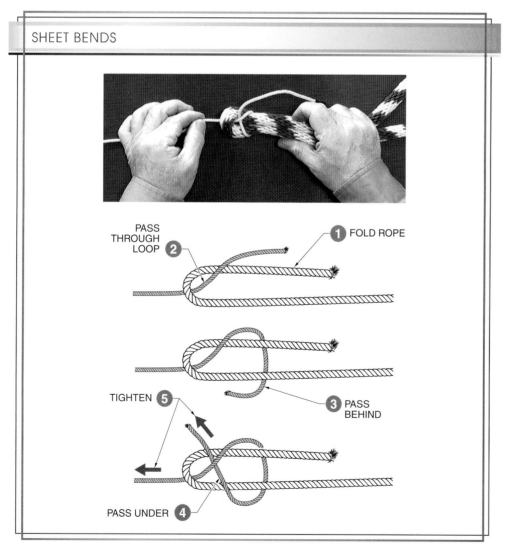

Figure 3-29. A sheet bend is an effective way to join two ropes temporarily.

For additional
information, visit
qr.njatcdb.org
Item #1667

Double Sheet Bends. The double sheet bend is a simple variation of the sheet bend that includes an additional round turn as a final step. **See Figure 3-30.** This is necessary when the two ropes are synthetic or of significantly different diameters, which increases the risk of spontaneous untying when under load. The additional turn helps lock the small diameter rope in place with pressure. A double sheet bend is tied by applying the following procedure:

1. Fold the thicker rope into a flat bight.
2. Pass the end of the thinner rope up through the bight's loop.
3. Pass the thinner rope around the thicker rope's folded working end and standing part.
4. Bring the thinner rope's end up to the front and then pass it underneath itself.
5. Pass the thinner rope's end around the knot again below the first turn.
6. Pass the thinner rope's end underneath itself again.
7. Tighten by pulling on the thinner rope.

DOUBLE SHEET BENDS

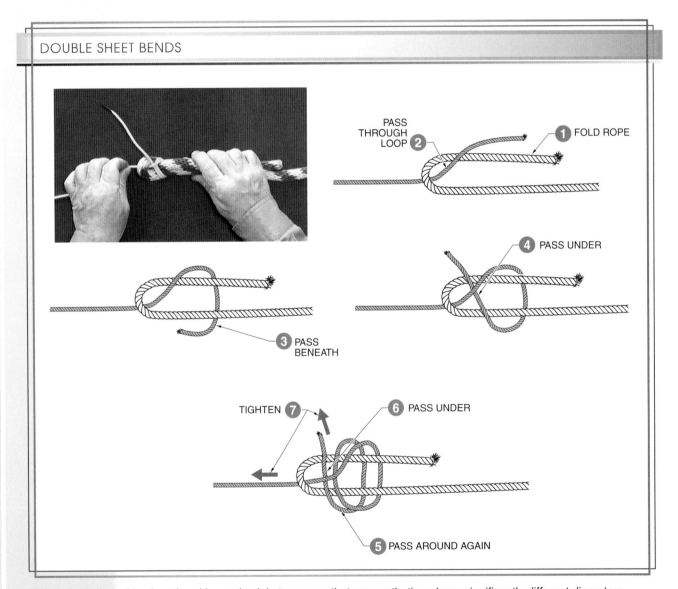

Figure 3-30. A double sheet bend is used to join two ropes that are synthetic or have significantly different diameters.

Common Rigging Hitches

Hitches rely on the pressure of ropes pressing together. The standing part of a rope is nipped, or jammed, over a working part. The friction of the nip causes the working part to be bound as the standing part is pulled, which prevents the working part from slipping through. These hitches allow a rigger to form a tight bind that is easily released after a lift. Due to the reliance on friction, hitches should never be formed with slippery rope, including some synthetic ropes.

The knot efficiencies of hitches can vary greatly, primarily because they are greatly affected by the size of the object being hitched. However, in most circumstances, they are roughly comparable to the efficiencies of most knots, averaging about 60%.

Half Hitches. A half hitch is the simplest knot of all. Indeed, it is also known as a simple hitch. By itself, a half hitch has little holding power, but it forms the basis for many other knots and hitches. It is simply a loop of rope around an object. The standing part nips, or presses, the working part against the object and holds via friction. **See Figure 3-31.** Multiple single hitches may be used to bundle lengths of tubing or lumber or to help stabilize a load as it is hoisted. Therefore, a half hitch relies on loading to maintain itself. If the standing part is unloaded, then the hitch loosens immediately. A half hitch on the bight is formed by applying the following procedure:

1. With the palm facing up, grasp the rope.

2. Twist the hand until the palm faces down, forming a loop in the rope.

3. Slide the loop over the object.

4. Tighten by pulling on the working end in the direction opposite the standing part.

For additional information, visit qr.njatcdb.org Item # 1668

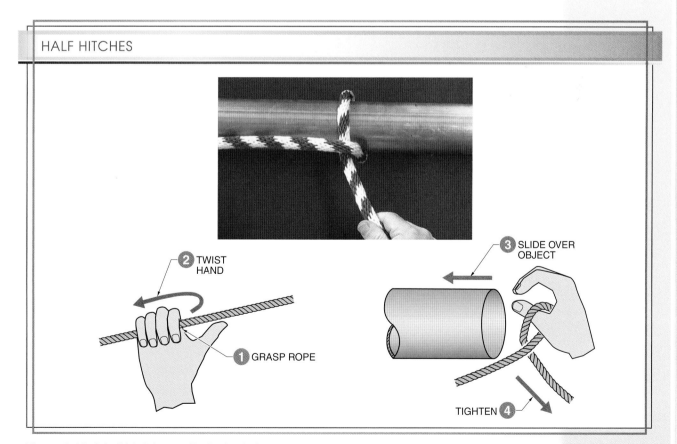

HALF HITCHES

② TWIST HAND

③ SLIDE OVER OBJECT

① GRASP ROPE

TIGHTEN **④**

Figure 3-31. A half hitch is usually the basis for a more complex knot or hitch.

Clove Hitches. A clove hitch is a combination of two half hitches in succession. **See Figure 3-32.** A clove hitch can be attached quickly and released rapidly. Tying a clove hitch is often part of a rigger exam, but the hitch is not very secure and should not be used to hoist a load unless both ends of the rope are kept under tension. A clove hitch should also never be used when tension may be applied perpendicular to a load. The tendency for the hitch to roll out if tension is released or if the load rotates can create an extremely dangerous situation.

Clove hitches can be formed either by threading the end or on the bight if the loops can be slipped over one end of an object. Therefore, there are two different ways to tie a clove hitch. A clove hitch by threading the end is formed by applying the following procedure:

1. Wrap the working end around the object.

2. Cross the standing part and wrap the working end around the object a second time.

3. Bring the working end up and then tuck it under the second turn.

4. Tighten by pulling on the working end.

A clove hitch on the bight is formed by applying the following procedure:

1. Form two loops of rope and hold them together in one hand.

2. Without turning the loops, move the front loop behind the back loop.

3. Slide the loops over the object.

4. Tighten by pulling on the working end.

CLOVE HITCHES

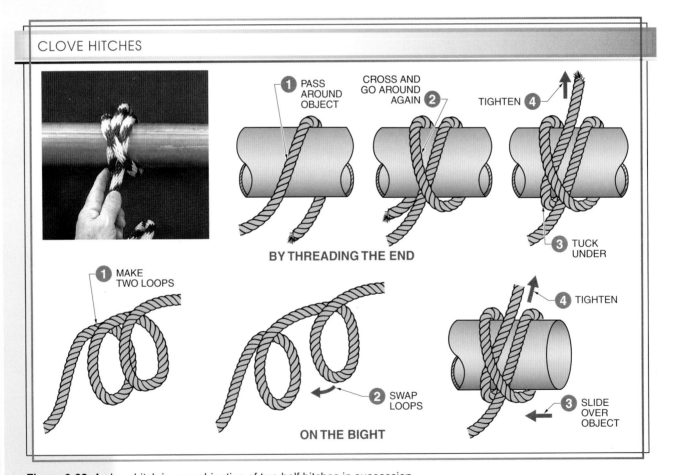

Figure 3-32. A clove hitch is a combination of two half hitches in succession.

Rolling Hitches. A rolling hitch is closely related to a clove hitch. However, a rolling hitch includes an additional turn and also supports loading from the side, almost parallel to the hitched object, as long as tension is applied to the standing part in the opposite direction of the working end. **See Figure 3-33.** When used in conjunction with one or two single hitches, the rolling hitch provides a means of hoisting a single piece of pipe or lumber vertically. The primary advantage of this hitch is that when the rope is slipped off of the object, all of the hitches fall out and the rope is ready to tie to another load. Like the clove hitch, the rolling hitch should not be used when tension will be applied perpendicular to the load, since it also has a tendency to roll out.

For additional information, visit qr.njatcdb.org Item #1670

ROLLING HITCHES

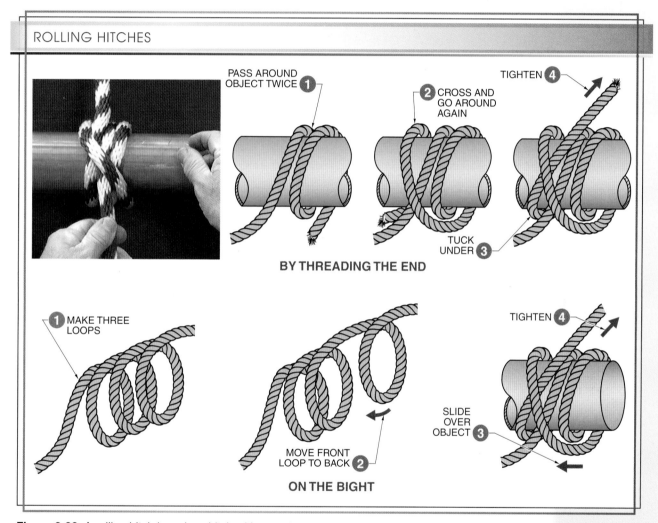

Figure 3-33. A rolling hitch is a clove hitch with an extra wrap.

If tied and used correctly, a rolling hitch is a secure knot. A rolling hitch by threading the end is formed by applying the following procedure:
1. Wrap the working end around the object twice.
2. Bring the end up and around the object again, this time crossing the other loops.
3. Bring the end up and then tuck under the last turn.
4. Tighten by pulling on the working end.

For additional
information, visit
qr.njatcdb.org
Item #1671

A rolling hitch on the bight is formed by applying the following procedure:

1. Form three loops of rope and hold them together in one hand.

2. Without turning the loops, move the front loop behind the back loop.

3. Slide the loops over the object.

4. Tighten by pulling on the working end.

Spar Hitches. A spar hitch is also closely related to the clove hitch. **See Figure 3-34.** The advantage of the spar hitch is that it is far less likely to roll out under tension that is either perpendicular or parallel to the load. However, unlike the clove and rolling hitches, when a spar hitch is slid off an object, it leaves a loose knot in the rope that must be untied. A spar hitch is formed by applying the following procedure:

1. Wrap the working end around the object.

2. Bring the end up and around the object again, this time crossing the first loop.

3. Bring the end up and over the last loop, and then tuck under the first loop.

4. Tighten by pulling on the working end.

SPAR HITCHES

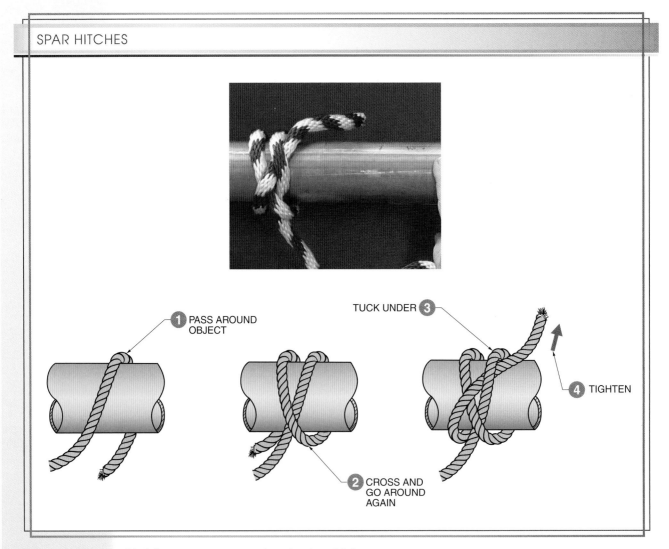

Figure 3-34. A spar hitch is a more secure version of a clove hitch.

Round Turns. A round turn is simply two loops of rope around an object. **See Figure 3-35.** It is easy to form under load. The friction of the loops of rope around the object provides some holding power. It is not enough to secure the load, but it helps prevent the rope from slipping. Additional knots, typically two half hitches, can then be tied to secure the round turn. A round turn is formed by applying the following procedure:

1. Wrap the working end around the object.
2. Bring the end up and around the object again.
3. Tighten by pulling on the working end.

For additional information, visit qr.njatcdb.org Item #1672

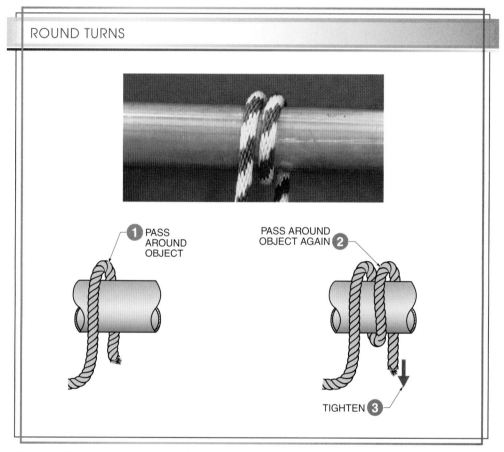

Figure 3-35. A round turn is simply two complete turns around an object.

Round Turn and Two Half Hitches. A round turn is typically finished with two half hitches, which form a clove hitch. **See Figure 3-36.** The half hitches are fastened to the standing part, not the object being hitched. The two basic components of this knot, the round turn and the half hitches, can be applied multiple times. For example, a double round turn with three half hitches is not uncommon. Additional round turns take up excess rope and provide extra friction. Additional half hitches help secure the knot further. A round turn and two half hitches are formed by applying the following procedure:

1. Wrap the working end around the object twice, forming a round turn.
2. Wrap the working end around the standing part to form a half hitch.

For additional information, visit qr.njatcdb.org Item #1673

3. Wrap the working end around the standing part again to form another half hitch, which also completes a clove hitch.

4. Tighten by pulling on the working end.

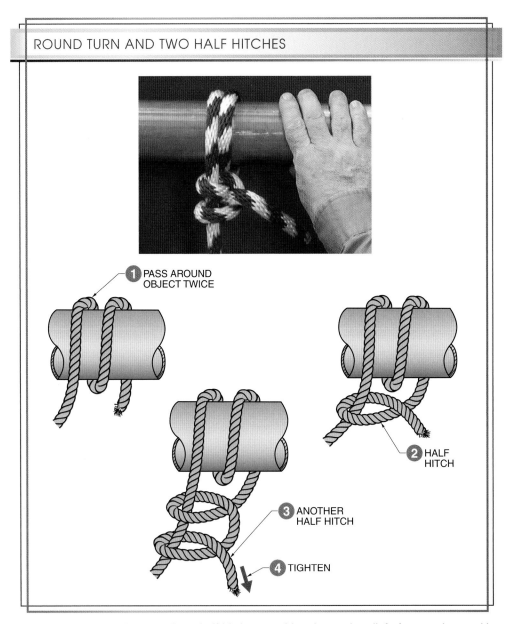

ROUND TURN AND TWO HALF HITCHES

1 PASS AROUND OBJECT TWICE

2 HALF HITCH

3 ANOTHER HALF HITCH

4 TIGHTEN

Figure 3-36. A round turn and two half hitches combine the strain relief of a round turn with the security of two half hitches.

Cow Hitches. The cow hitch is closely related to the clove hitch. Both are composed of a pair of half hitches, though the half hitches face the same direction in a clove hitch and face opposite directions in a cow hitch. **See Figure 3-37.** A cow hitch is often used to quickly attach a tag line to a load. It is made and released easily but is firm enough to steady loads as long as tension is applied equally to both ends. If loaded unequally, the rope can run through the hitch fairly easily. Therefore, this hitch is typically not recommended for critical applications. A cow hitch is formed by applying the following procedure:

1. Fold the rope and hold the bight.
2. Open the bight at the fold and turn it over onto the ropes.
3. Grasp the ropes through the loop and pull the loop over them.
4. Hold the two resulting loops open and slide them over the object.
5. Tighten by pulling on both rope ends.

COW HITCHES

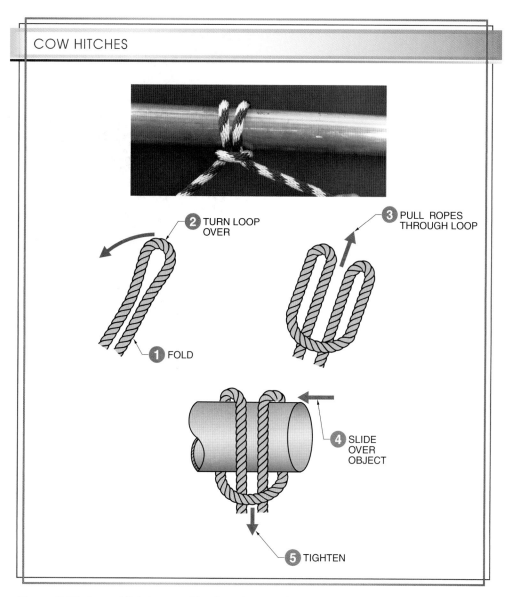

Figure 3-37. A cow hitch is a combination of two half hitches facing opposite directions.

For additional information, visit qr.njatcdb.org Item #1675

Cat's Paw Hitches. A cat's paw hitch is used as a quickly formed, light-duty eye that can be hung on a hook. **See Figure 3-38.** This hitch is similar to a cow hitch, but the additional twists on each side add stability and prevent the rope from sliding through the hook due to unequal tension from an asymmetrical load. The number of twists can vary, but typically three or four are sufficient. A cat's paw hitch is formed by applying the following procedure:

1. Grasp the rope with both hands. Leave plenty of bight.

2. Rotate hands in opposite directions to form two loops.

3. Continue to rotate the two loops for two or more complete turns.

4. Place the loops together over the end of a hook.

5. Tighten by pulling on both rope ends.

CAT'S PAW HITCHES

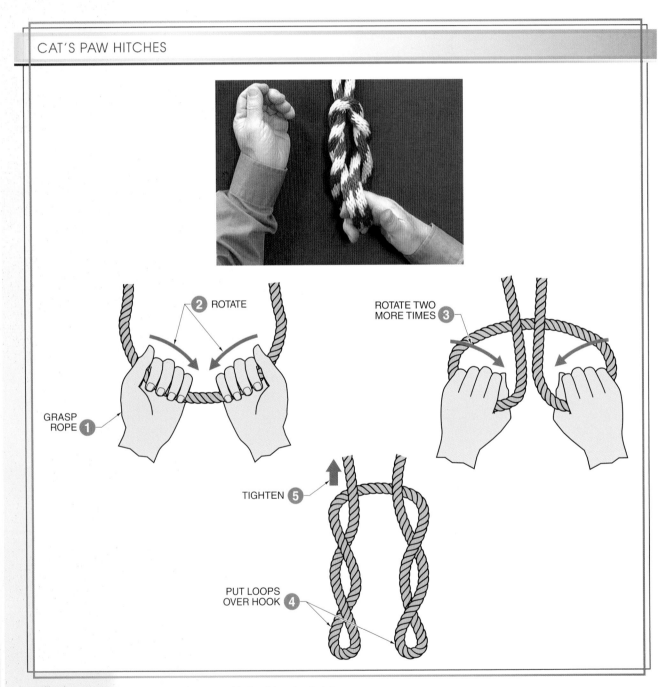

Figure 3-38. A cat's paw hitch is a cow hitch with extra twists.

Trucker's Hitches. A trucker's hitch is a compound knot that has the distinct feature of providing mechanical advantage when tightening. There are many variations, though the hitch basically consists of a loop knot that is formed at one end or on the bight, the rope running down through a tie-down point and back up through the loop, and the rope then being pulled tight and secured with another knot. **See Figure 3-39.** This arrangement provides a mechanical advantage, meaning that the force used to tighten the knot is multiplied into a greater force that binds the load.

The individual knots used to form the loop and secure the working end may vary. A common combination is a lineman's loop with a pair of half hitches. This trucker's hitch is formed by applying the following procedure:

1. Tie a lineman's loop near the working end of the rope.
2. Pass the working end down through a tie-down point and back up toward the loop.
3. Pass the working end through the loop.
4. Tighten by pulling down on the working end.
5. Secure the working end to the standing end below the loop with two half hitches.

For additional information, visit qr.njatcdb.org Item #1676

TRUCKER'S HITCHES

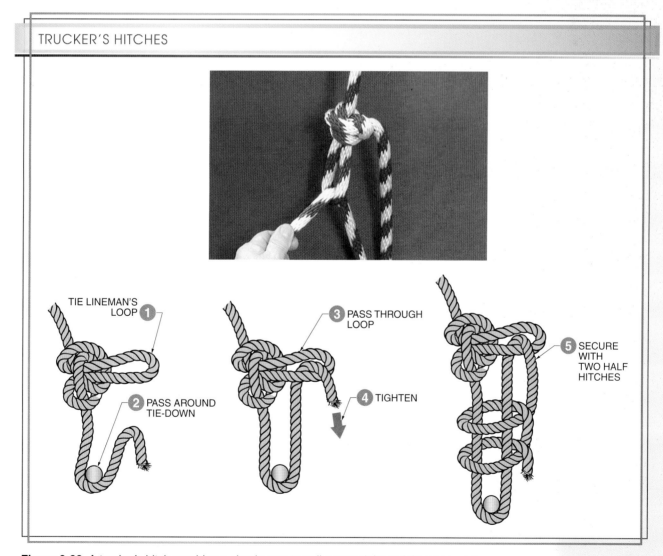

Figure 3-39. A trucker's hitch provides a simple way to pull a rope tight and then secure it.

Timber Hitches. A timber hitch is a knot that is used to pull an object, such as a utility pole. **See Figure 3-40.** A timber hitch is formed by applying the following procedure:

1. Pass the working end around the cylindrical object.
2. Pass the working end around the standing part.
3. Wrap the working end around itself a minimum of four times.
4. Tighten by pulling on the working end.

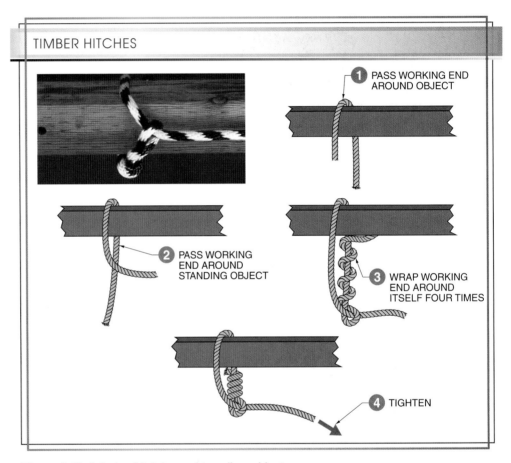

Figure 3-40. A timber hitch is used to pull an object.

Munter Hitches. A munter hitch is a knot that uses friction for repelling or belaying by having the rope rub against itself and the object it has been wrapped around. The munter hitch is typically wrapped around a carabiner. A munter hitch is formed by applying the following procedure:

1. Form a loop.
2. Fold one standing part back on the loop.
3. Clip a carabiner around the loop and standing part.

ROPE APPLICATIONS

Bull rope is heavy rope that is used to pull wires. It is typically an inexpensive, large diameter rope used for noncritical loads. A *hand line* is a rope, block, and set of snaps or hooks that can be attached to the crossarm of a utility pole. It is used to hoist tools and equipment while on the utility pole. A hand line may also be used in an emergency to rescue a lineman from a utility pole top. A *tag line* is a rope, handled by a qualified individual, used to control rotational movement of a load during a lift. *Pulling rope* is a high-strength, low-stretch rope that is used for tension stringing conductors. Pulling rope does not stretch much when pulling a long section of wire.

LASHING

Lashing is a method of binding two objects to each other by wrapping rope around them numerous times. Typically, two utility poles are lashed to each other or a rope block is lashed to a utility pole. **See Figure 3-41.** Utility poles are lashed together by applying the following procedure:

1. Form a rolling hitch around one pole.
2. Wrap both poles several times based on rope diameter.
3. Pass the working end between the two poles and make two or three passes around the horizontal wraps.
4. Form another rolling hitch on the other pole diagonal to the first hitch.

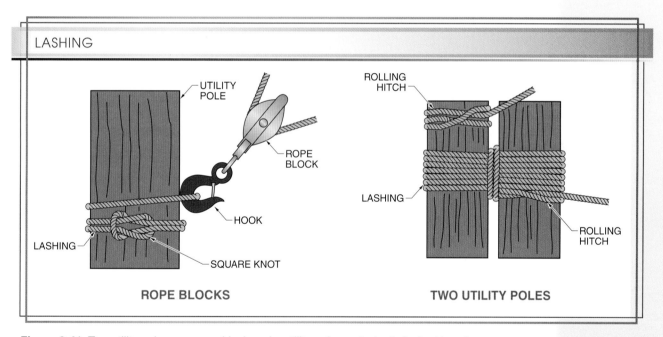

Figure 3-41. Two utility poles or a rope block and a utility pole are typically lashed together.

A *rope block* is an assembly of one or more sheaves in a frame. A sheave, also known as a pulley, is a grooved wheel attached to a frame or block that supports a rope that is changing direction. Rope blocks are lashed to utility poles for a 90° pulling angle by applying the following procedure:

1. Wrap a rope three or four times around a pole with the ends at the bottom of the wraps.
2. Tie a square knot at the bottom of the wraps.
3. Attach a rope block to the top wrap of rope with the back of the hook against the pole.

Rope blocks are lashed to utility poles for a downward pulling angle by applying the following procedure:

1. Wrap a rope three or four times around a pole with the ends at the top of the wraps.

2. Tie a square knot at the top of the wraps.

3. Attach a rope block to the bottom wrap of rope with the back of the hook against the pole.

STAKE HOLDFASTS

Guy wires are used to stabilize utility poles and offset the lateral pull of overhead conductors. Stake holdfasts are used as temporary anchors when using temporary pole guys. Two or more stakes can be driven into the ground and used as an anchor for guy lines. **See Figure 3-42.** The stakes may be steel digging bars or steel stakes. A stake holdfast is made by applying the following procedure:

1. Drive a stake into the ground at a 90° angle to the guy wire.

2. Drive another stake into the ground several inches away from the first stake and at the same 90° angle to the guy wire.

3. Form a loop around the stakes and tie it with a square knot.

4. Insert a smaller stake or piece of wood in the loop and twist until it touches the ground and is locked in place.

5. Form a round turn and two half hitches to secure the guy wire to the first stake and utility pole.

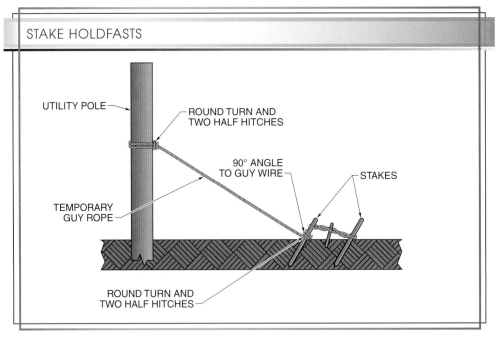

STAKE HOLDFASTS

UTILITY POLE

ROUND TURN AND
TWO HALF HITCHES

90° ANGLE
TO GUY WIRE

STAKES

TEMPORARY
GUY ROPE

ROUND TURN AND
TWO HALF HITCHES

Figure 3-42. Stake holdfasts are used as temporary anchors for temporary pole guys.

COMMON APPLICATIONS— STRINGING OVERHEAD CONDUCTORS

Electric power distribution systems contain overhead conductors that are strung for miles in order to distribute electric power. Overhead conductors are typically strung through a number of utility poles to safely deliver power. A conductor can be hoisted and pulled through a series of utility poles by attaching a pulling grip to the end of the conductor and using a puller and bullwheel tensioner to string the conductor through a series of stringing blocks attached to each utility pole.

Brooks Brothers Trailers

To string an overhead conductor, apply the following procedure:

1. Set up a bullwheel tensioner near the first utility pole and a puller near the last utility pole where the conductor is to be strung.
2. Place the conductor reel on the attachment behind the bullwheel tensioner.
3. String a pulling line, typically steel cable or synthetic rope, from the puller through each stringing block to the bullwheel tensioner.
4. Wrap the pulling line halfway around the bullwheel tensioner and attach the pulling line to the conductor using the proper pulling grip.
5. Pull the conductor through each stringing block while maintaining tension between the bullwheel tensioner and the puller to keep the conductor clear of obstructions and from sagging too much.

Chapter 3 Learner Resources

For additional information, visit
qr.njatcdb.org
Item # 1649

Lineworker Rigging Practices

WIRE ROPE AND CHAIN 4

Lift-All Company, Inc.

OBJECTIVES

- Describe the basic construction of wire rope.
- Compare the types of lay and strand patterns for wire rope and their effect on a wire rope's characteristics.
- Estimate the rated load of wire rope, accounting for safety factor and bending efficiency.
- Identify common types of wire rope wear and damage.
- Compare the advantages and disadvantages of chain for lifting purposes.
- Describe the materials and methods used to construct chain.
- Specify how extreme temperature affects chain strength.
- List the inspection criteria for removing chain or chain slings from service.

Wire rope is commonly used as a lifting line in hoists and cranes because it is easily used with sheaves (pulleys) and wound onto drums. Wire rope is strong and durable. Chain is a series of interlocking metal links. It is commonly used for lifting applications due to its high strength and extreme durability, even in extreme environments. It also has the advantage of being highly flexible, which allows it to be wrapped around some loads. However, chain should not be used with delicate loads as it can cause damage and abrasion on load surfaces.

WIRE ROPE

Rope is a length of fibers or thin wires that are twisted or braided together to form a strong and flexible line. **See Figure 4-1.** Most rope used in rigging and hoisting applications is wire rope. Wire rope is a highly specialized precision product that is adaptable to many uses and conditions of operation. To meet the requirements of different types of service, ropes are designed and manufactured in a number of constructions and grades. The appropriate wire rope is selected based on the load weight, potential for shock from acceleration or deceleration, attachments needed, and lifting conditions.

WIRE ROPE

Figure 4-1. Wire rope is used with hoists and cranes and can also be used to make lifting slings.

Wire Metals

Rope wires can be made from several types of metal, including steel, iron, stainless steel, Monel®, and bronze. Wire rope manufacturers select the wire that is most appropriate for the requirements of the rope. The most widely used material is high-carbon steel, which is available in a variety of grades, each providing slightly different properties to the wire. Grades of wire rope steel include traction steel (TS), mild plow steel (MPS), plow steel (PS), improved plow steel (IPS), and extra improved plow steel (EIPS).

Typical rigging and lifting wire rope is bright (uncoated). Steel that is galvanized (coated with zinc) may be required for harsh or corrosive environments. Galvanized wire rope is approximately 10% lower in strength than bright wire rope.

Wire Rope Construction

Wire rope is manufactured from thin metal wires. A certain number of metal wires are twisted into strands. A *strand* is a bundle of wires twisted spirally around an axis. Then a certain number of strands are wound spirally around a core. **See Figure 4-2.** The *core* is the strand of metal wire or fiber that forms the center of a wire rope. The strength and flexibility of a rope depends on the precise laying of each wire and the way the wires slide against each other as the rope flexes.

WIRE ROPE CONSTRUCTION

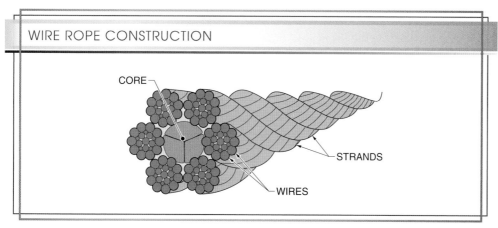

Figure 4-2. Wire rope is composed of a core surrounded by strands, each of which is composed of a specific pattern of wires of different sizes.

Lay. The *lay* is a designation for the direction in which rope strands are twisted, described as if spiraling away from the observer. **See Figure 4-3.** *Right-hand lay rope* is rope with strands that spiral to the right (clockwise). *Left-hand lay rope* is rope with strands that spiral to the left (counterclockwise). The lay of the strands, in combination with the twist direction of the wires, results in regular-lay or lang-lay rope.

ROPE LAY

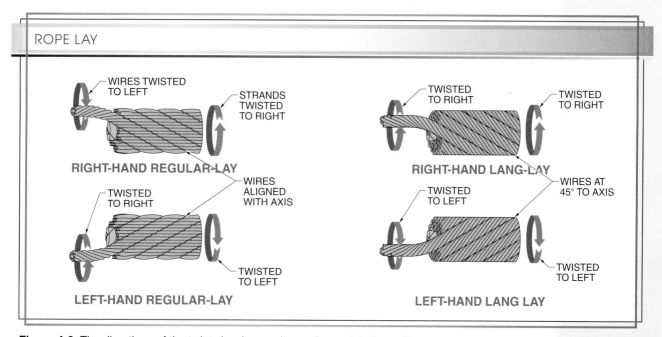

Figure 4-3. The directions of the twists in wires and strands result in four different rope twist combinations.

A *regular-lay rope* is rope in which the wires in the strands are twisted in the opposite direction of the lay of the strands. In a right-hand regular-lay rope, the strands are twisted to the right, and the wires are twisted to the left. Similarly, in a left-hand regular-lay rope, the strands are twisted to the left, and the wires are twisted to the right. Regular-lay rope is easily identified because the wires follow the direction of the axis of the rope. Regular-lay rope resists untwisting under load.

A *lang-lay rope* is rope in which the wires and strands are twisted in the same direction. A right-hand lang-lay rope has wires that are twisted to the right and strands that are twisted to the right. A left-hand lang-lay rope has wires that are twisted to the left and strands that are twisted to the left. Lang-lay ropes are quickly identified by their wires crossing the axis of the rope at approximately 45°. Lang-lay ropes expose more wires to wear at the outer surface, which increases the useful life of the ropes. Lang-lay ropes also have greater flexibility than regular-lay ropes but have a tendency to untwist under tension.

Since it is constructed from twisted strands, rope may spin or rotate when used. *Cabling* is the tendency of a rope to rotate and untwist when under load. This must not be allowed because it transfers all the load weight to the shorter, center core. Rotation-resistant ropes that consist of two or more layers of strands twisting in opposing directions are available to counteract the cabling tendency. These ropes require special handling because they are susceptible to kinking, crushing, and unbalancing. They also tend to have lower breaking strengths than similarly sized ropes.

Strand Patterns. Wire rope strands often have multiple sizes of wire, arranged in specific patterns, to provide desired flexibility and wear characteristics for the rope. **See Figure 4-4.** While many patterns exist, the most common designs are filler wire, Warrington, Seale, and Warrington-Seale.

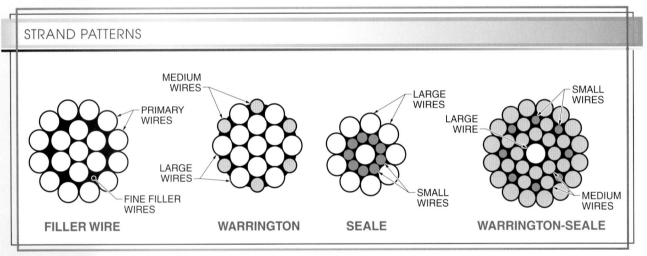

STRAND PATTERNS

Figure 4-4. The wires in strands can be arranged in different patterns, which changes the strength, flexibility, and wear characteristics of the rope.

Filler wire strands have fine wires that fill the gaps between the major wires. The fine wires provide stability to the shape of the strand but contribute little strength. Filler wire rope is the most flexible but wears more quickly than Warrington or Seale wire rope.

Warrington wire strands are constructed of two or more wire sizes. The Warrington pattern is easily distinguished by its alternating large and small wire sizes in the outer layer. The wire sizes in the interior layers vary, depending on the strand size. Warrington wire rope is less flexible than filler wire rope but has better wear resistance.

Seale wire strands also have multiple wire sizes but are not mixed within individual layers. All wires in the outermost layer are one size, and all wires in the next layer are another size. Seale wire rope is less flexible than Warrington wire rope but is the least susceptible to wear.

Combination Warrington-Seale wire strands possess the qualities of both wear resistance and flexibility. The outer layer is typically a Seale pattern, used for increased wear resistance. An inner layer is a Warrington wire pattern that provides increased flexibility. Warrington and Seale patterns may be alternated for each layer, and filler wires may be added between them.

Cores. Wire rope strands are laid around a core. The core may be a small wire rope, wire strand, or fiber (nonmetallic) strand.

An independent wire rope core (IWRC) is itself a small wire rope, made from smaller wires. Wire ropes with an IWRC resist crushing and have consistent stretching properties.

A wire strand core (WSC) is a single strand constructed of multiple wires. This core strand is typically of the same size and pattern as the rope strands.

A fiber core (FC) is made of synthetic or natural fibers, either polypropylene or hemp/sisal, and is shaped to keep the strands in place and cushioned. Fiber cores contribute greater flexibility to wire rope but are not as strong as WSC or IWRC types.

Designations. Wire rope is classified according to the number of strands in the rope and the number of wires in each strand. For example, a 6 × 7 rope indicates a six-stranded rope with approximately 7 wires per strand. The second number in the designation is nominal in that the number of wires in a strand may be slightly higher or lower. For example, a 6 × 7 rope has between 3 and 14 wires per strand. Similarly, a 6 × 19 rope has between 15 and 26 wires per strand. The 6 × 19 classification includes constructions such as 6 × 21 filler wire, 6 × 25 filler wire, and 6 × 26 Warrington-Seale. These constructions are all classified as 6 × 19 rope, despite the fact that none of their strands have exactly 19 wires.

Acronyms are added to rope designations to indicate the rope construction, wire material, core material, core construction, strand pattern, and other specifications. **See Figure 4-5.** For example, the designation "6 × 19W+FC RH OL FSWR" indicates a six-strand flexible steel wire rope with approximately 19 wires per strand in a Warrington pattern, with a fiber core and right-hand ordinary (regular) lay.

ROPE DESIGNATION ACRONYMS

Rope Construction		Materials	
RH	Right-hand lay	TS	Traction steel
LH	Left-hand lay	MPS	Mild plow steel
RL	Regular lay	PS	Plow steel
OL	Ordinary (regular) lay	IPS	Improved plow steel
LL	Lang lay	GIPS	Galvanized improved plow steel
AL	Alternate lay	EIPS	Extra improved plow steel
RR	Rotation-resistant	FSWR	Flexible steel wire rope
NR	Nonrotating	J	Jute (fiber)
Strand Patterns		Cores	
FW	Filler wire	FC	Fiber core
S	Seale	HC	Sisal core
SF	Seale filler wire	WSC	Wire strand core
W	Warrington	IWS	Independent wire strand
SW	Seale-Warrington	WRC	Wire rope core
WS	Warrington-Seale	IWR	Independent wire rope
TS	Triangular strand	IWRC	Independent wire rope core

Figure 4-5. Acronyms are used for rope designations to indicate rope characteristics.

Wire Rope Diameter

Due to the construction of rope, its cross section is not a perfect circle. Measuring the diameter of a wire rope accurately is problematic because it is possible to find different measurements based on the placement of the measuring calipers. Diameter measurements must be precise in order to detect excessive stretching in the rope that warrants its removal from service. Therefore, the diameter of wire rope has been defined as the smallest possible dimension that fully encircles the rope. **See Figure 4-6.** To measure rope diameter, the distance from a high spot on one side of the rope to a high spot on the opposite side of the rope is measured using calipers.

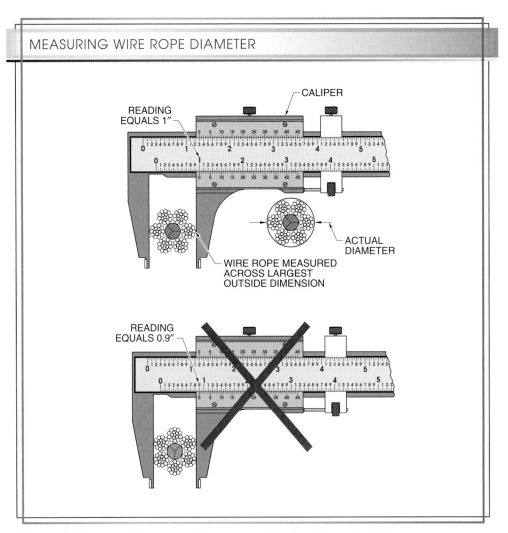

MEASURING WIRE ROPE DIAMETER

Figure 4-6. Wire rope diameter must be measured at the widest point of the rope.

Wire Rope Strength

The strength rating that is specified for rope is typically its breaking strength, not its rated load. **See Figure 4-7.** This is because there are many possible uses for wire rope, and each application may have its own safety factor. Therefore, the hoisting personnel must carefully check the rope ratings and, if necessary, apply the appropriate factors and deratings in order to determine the appropriate rated load for the application. Safety factor, bending efficiency, and environmental conditions all affect rope strength.

BREAKING STRENGTHS OF SELECTED WIRE ROPES

Diameter[†]	Improved Plow Steel*		Extra-Improved Plow Steel*
	Fiber Core	IWRC[†]	IWRC[†]
1/4	5340	5740	6640
5/16	8300	8940	10,280
3/8	11,900	12,800	14,720
7/16	16,120	17,340	19,900
1/2	20,800	22,400	26,000
9/16	26,400	28,200	32,800
5/8	32,600	35,000	40,200
3/4	46,400	50,000	57,400
7/8	62,800	67,400	77,600
1	81,600	87,600	100,800

* in lb, for uncoated, general purpose, rotation-resistant 6 × 19 (class 2) or
  6 × 37 (class 3) wire rope
† in in.
‡ independent wire rope core

Figure 4-7. The breaking strengths of wire ropes vary by type, material, and diameter.

Rated Loads. A safety factor is used to determine the rated load from the breaking strength. Most rigging applications use a safety factor of 5, meaning that the rated load is one-fifth of the breaking strength. This provides a large margin of safety against failure due to normal variances in conditions of use.

A rope is typically selected based on the maximum load it must support. The minimum breaking strength required to support a certain rated load is calculated using the following formula:

$$S_{break} = RL \times f_{safety}$$

where

S_{break} = breaking strength (in lb)

RL = rated load (in lb)

f_{safety} = safety factor

For example, what is the minimum wire rope breaking strength required to lift a 4000 lb milling machine directly with a hoist? *Note:* A safety factor of 5 is used because the load is a steady lift without shock.

$$S_{break} = RL \times f_{safety}$$
$$S_{break} = 4000 \times 5$$
$$S_{break} = \textbf{20,000 lb}$$

This information would then be used to choose an appropriate rope or confirm that an existing rope is adequate for the lift. In this example, the machine can be safely hoisted with a 1/2″ (or larger) wire rope of any of the common types.

Bending Efficiency. Ropes are often wrapped over sheaves (pulleys) or around loads while being used. This bending puts a rope under additional mechanical stress, which reduces its ability to withstand tension forces. **See Figure 4-8.** In fact, bending a rope over a small diameter can reduce its effective strength by more than 50%.

The degree to which a rope's effective strength is reduced due to bending depends on the bend ratio. The *bend ratio* is the ratio of the diameter of a bend to the nominal diameter of the rope. The bend ratio is also known as the D/d ratio and is calculated using the following formula:

$$R_{bend} = \frac{D}{d}$$

where

R_{bend} = bend ratio

D = diameter of rope bend (in in.)

d = diameter of rope (in in.)

ROPE BENDING

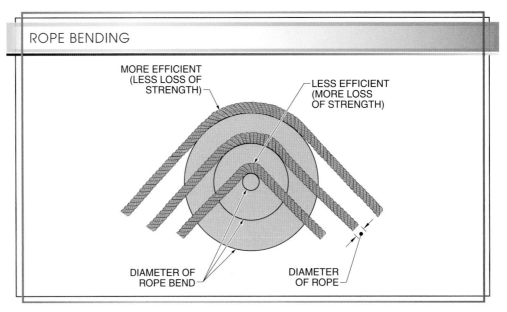

MORE EFFICIENT
(LESS LOSS OF
STRENGTH)

LESS EFFICIENT
(MORE LOSS
OF STRENGTH)

DIAMETER OF
ROPE BEND

DIAMETER
OF ROPE

Figure 4-8. Tight bends reduce the effective strength of a rope and can permanently damage the wires.

The calculated bend ratio is then used with a chart or plot to determine the rope bending efficiency. *Bending efficiency* is the ratio of the strength of a bent rope to its nominal strength rating. **See Figure 4-9.** This efficiency data is compiled by laboratories that conduct load tests on ropes. The efficiency percentage is then used to derate the rope's rated load using the following formula:

$$RL_{bend} = RL \times h_{bend}$$

where

RL_{bend} = rated load of rope after bending (in lb)

RL = rated load of rope (in lb)

h_{bend} = bending efficiency

For example, what is the rated load after bending a ⅜″ (0.375″) rope, which has a rated load of 2380 lb, (11,900 lb divided by a safety factor of 5) when wrapped over a 6″ diameter sheave? First, calculate the bend ratio.

$$R_{bend} = \frac{D}{d}$$

$$R_{bend} = \frac{6}{0.375}$$

$$R_{bend} = \mathbf{16}$$

The bend ratio of 16 is then used with a chart or plot for that particular rope to determine the resulting bending efficiency. (Note that while the shapes of bending efficiency curves are similar, the values may vary between different ropes. Manufacturer's specifications should always be consulted when derating a rope for bending efficiency.) In this example, the rope has a bending efficiency of approximately 90% (0.90).

$$RL_{bend} = RL \times h_{bend}$$
$$RL_{bend} = 2380 \times 0.90$$
$$RL_{bend} = \textbf{2142 lb}$$

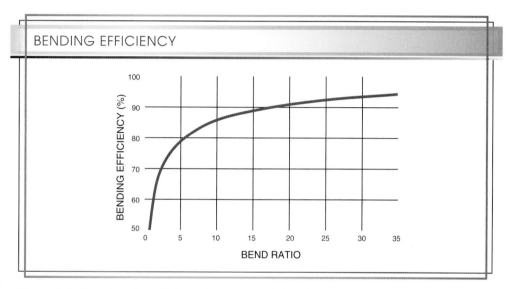

BENDING EFFICIENCY

Figure 4-9. Rope bending efficiency increases with bend ratio.

Therefore, even though this rope has a rated load of 2380 lb, since it is used in a way that bends the rope, it should not be subjected to forces greater than 2142 lb. If the rated load ends up being too low for the application, either a different rope must be used or the rigging changed to reduce bending. For example, a larger sheave can be used.

Moisture Exposure. The effects of moisture vary between rope types. Moisture causes steel wire rope to rust, which is a type of corrosion. *Corrosion* is the disintegration of a material due to chemical reaction with its environment. Since corrosion may occur on the inside first, damage due to rust may not be visible before the rope breaks. Wire ropes should be kept well lubricated to prevent rusting.

Extreme Temperatures. Manufacturers supply data on the temperature limits of rope. Wire rope with a fiber core is typically rated for temperatures up to 180°F (82°C). Wire rope with a wire core is typically rated for temperatures up to 400°F (204°C). Overheating wire rope destroys the lubrication applied during manufacturing, eventually leading to corrosion and loss of flexibility. Severe overheating may directly affect the metal, immediately weakening the rope.

Chemical Activity. Exposure to corrosive chemicals can cause significant and rapid damage to rope. Rope used in chemically corrosive environments, such as battery shops, metal-plating shops, pickling plants, or pulp and paper mills, must be designed specifically to resist the chemicals present. Wire rope made from stainless steel or coated with vinyl, nylon, Teflon®, or zinc is recommended for corrosion resistance.

WIRE ROPE INSPECTION

After a wire rope has been in service a short while, its breaking strength actually increases slightly. **See Figure 4-10.** This is because the wires settle into position within the strands, making the strands more solid and uniform. However, further normal use then decreases the strength of the wire rope, first gradually and then rapidly, due to wire breaks, corrosion, and abrasion. This process can be monitored with frequent inspections so a wire rope can be removed from service before the strength decreases significantly.

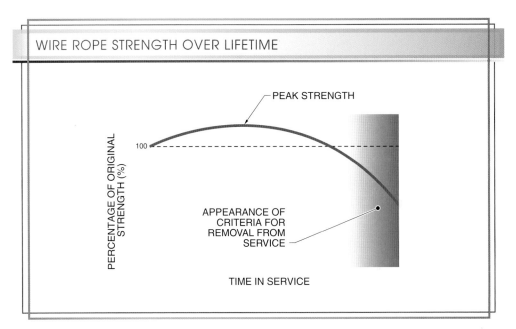

WIRE ROPE STRENGTH OVER LIFETIME

PEAK STRENGTH

100

PERCENTAGE OF ORIGINAL STRENGTH (%)

APPEARANCE OF CRITERIA FOR REMOVAL FROM SERVICE

TIME IN SERVICE

Figure 4-10. Wire rope strengthens slightly after use and then begins to lose strength. It should be removed from service before it is significantly weakened.

Broken wires are a common problem. **See Figure 4-11.** Damage in a small area indicates physical contact with sharp edges. Wires broken in many places throughout the rope indicate a severe overloading condition. A rope can tolerate a small number of broken wires, but a rope must be removed from service if it has an excessive number of wire breaks. The limit to the number of allowable wire breaks may vary depending on the rope size and type. However, a rope with a break in an entire strand must be removed immediately.

Corrosion may be general or localized. Signs of corrosion include discolored wires, rusty residue, a roughened or pitted surface, or slackness within strands. A loss of diameter, often caused by corrosion, of more than 10% is considered reason for removal of the rope from service.

Kinking is a sharp bend that permanently deforms the lay of rope strands. Kinking is caused by the tightening of a loop that is restrained from untwisting. This can happen from improper storage or improper removal of wire rope from a spool. Kinking significantly weakens a wire rope.

Wire rope can be crushed when trapped under or between heavy loads. The pressure distorts the arrangement of the wires in the strands and the strands in the rope. It can also break individual wires. Any sign of crushing requires removal of the rope from service.

Excessive wear on a wire rope flattens the outer layer of wires. Wear is quantified by periodically measuring the rope diameter. Rope should be replaced when the diameter narrows beyond the allowable limit. **See Figure 4-12.**

TYPES OF WIRE ROPE DAMAGE

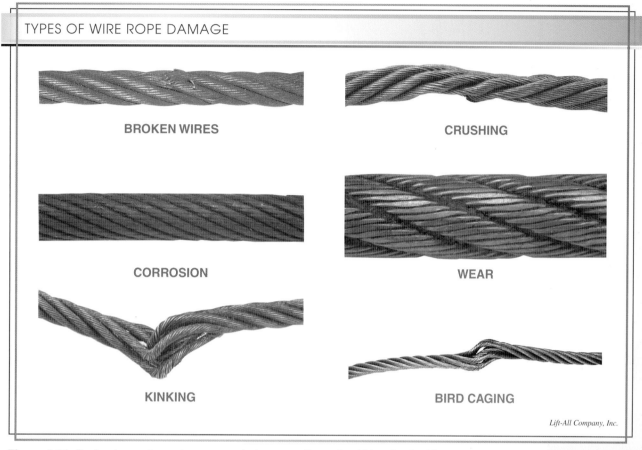

BROKEN WIRES

CRUSHING

CORROSION

WEAR

KINKING

BIRD CAGING

Lift-All Company, Inc.

Figure 4-11. During inspection, wire rope and wire rope slings should be checked for evidence of excessive wear or other types of damage.

WIRE ROPE WEAR LIMITS

Nominal Rope Diameter*	Allowable Reduction*
Up to 5/16	1/64
3/8 to 1/2	1/32
9/16 to 3/4	3/64
7/8 to 1 1/8	1/16
Over 1 1/8	3/32

* in in.

Figure 4-12. Wire rope should be replaced when wear reduces its diameter by a certain amount.

Bird caging is a type of damage to wire rope in which the outer strands separate and open. This is caused when an outer layer of strands becomes longer than an inner layer. Bird caging occurs from tight sheaves, shock loading, incorrect fitting installation or swivel use, or the application of a heavy load before the strands have settled.

Loose outer wires can form small loops, often in multiple strands along one side of a rope. If rope use continues, the loops can become flattened into small tangles of wires. This deformation is often caused by shock loading, tight bends, or kinks.

Core protrusion occurs when strands are forced apart and the core material is squeezed out from between the strands. This can be caused by shock loading or, in the case of fiber cores, swelling due to the absorption of moisture. This protrusion is also called a node, due to the localized increase in rope diameter.

CHAIN

A *chain* is a series of connected metal links. **See Figure 4-13.** Chain is recommended for rugged industrial applications where flexibility, abrasion resistance, and long life are required. Chain can often be used in situations in which other lifting materials would be damaged by the load, such as from rough or raw castings, or in which the materials would be damaged by environmental conditions, such as extreme temperatures. Chain is also widely used in construction work involving excavation, where dirt, moisture, and abrasion would quickly destroy other types of slings.

CHAIN

Figure 4-13. Chain is particularly useful for rigging rough-surfaced loads or rigging in harsh environments because it is extremely durable.

The use of chain for rigging is often favored over wire rope because chain has approximately three times the impact-absorption capability and is more flexible. Also, wire rope costs more than chain of similar strength and has only 5% of the expected service life.

Chain slings are available with many of the same types of hardware components available for wire rope slings, so they can often be used for similar applications.

Since both rope and webbing are composed of many load-bearing elements in parallel, they can tolerate a certain degree of failure from some elements while still providing significant strength capacity. For example, a wire rope could have some frayed wires and still be safe to use. **See Figure 4-14.** Chain relies on load-bearing elements arranged in a series. The failure of any single link may lead to the failure of the entire chain. This is a disadvantage unique to chain.

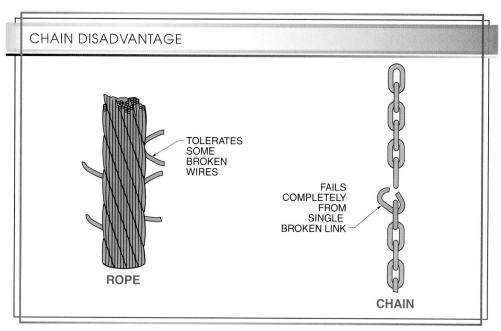

CHAIN DISADVANTAGE

TOLERATES SOME BROKEN WIRES

FAILS COMPLETELY FROM SINGLE BROKEN LINK

ROPE

CHAIN

Figure 4-14. A disadvantage of chain is that it can fail completely from the failure of a single element (link), while rope can sustain some frayed wires and still be safe to use.

The National Association of Chain Manufacturers (NACM), in conjunction with the ISO, standardizes the material and manufacturing specifications for chain. The standard ASME B30.9, *Slings*, includes information about the use of chain for lifting applications.

Chain Material

The strength of chain and chain attachments depend on the steel alloys from which they are made. An *alloy* is a metal formulated from the combination of two or more elements. The alloy composition and heat-treating processes used to form the alloy determine the metal's strength, hardness, and other characteristics.

Of the many types of steel alloy chain, Grade 80 and Grade 100 chain are most commonly specified for sling or tie-down applications. Other alloys may be used, but the manufacturer must provide specific data regarding the load rating and conditions of use. Nonalloy chain should never be used for lifting purposes.

If a chain is needed in a chemically active environment, it may need to be made of a special non-corroding alloy. The conditions of use should be evaluated by a qualified person and the chain material chosen accordingly. For example, stainless steel alloy chains are available and can be used under conditions in which regular steel alloy chains would be adversely affected by chemicals.

Chain Construction

Lifting chain is formed from steel rod cut into short lengths. Each piece is held against a forming die while rollers bend the rod into a link shape, with the ends meeting on a long side. The forming machine then welds the ends together as the next link is being formed. This weld can be seen as a slight bulge around the rod on one side of a link. **See Figure 4-15.** The next piece of rod is inserted through the link, rolled into the same shape, and welded. This process is continued until the desired length of chain is constructed.

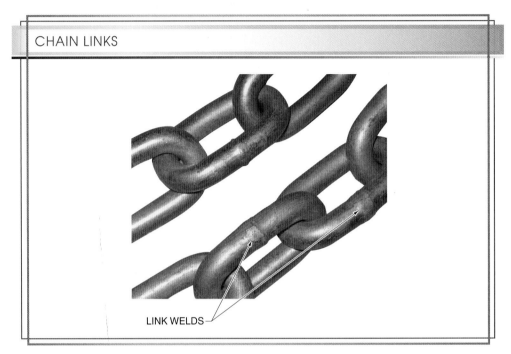

CHAIN LINKS

LINK WELDS

Figure 4-15. Chain is manufactured by bending short lengths of steel rod into links and welding the ends together.

The completed chain may then be further processed. Heat treatment can be used to develop the desired combination of strength and durability in the metal. Surface treatments may also be applied for a particular surface finish or for corrosion resistance.

Chain Strength

For a particular grade, the rated loads are based on nominal chain size, which is the approximate diameter of the rod used to form the links. **See Figure 4-16.** The standard safety factor for chain slings is four.

Chain is particularly suited for high-temperature applications in which other types of sling materials cannot be used. However, extreme temperatures can still affect a chain's rated loads. **See Figure 4-17.** High temperatures soften the metal, reducing the rated load, sometimes permanently. Conversely, extremely low temperatures make the metal brittle and more likely to fracture.

According to ASME B30.9, the normal temperature range for chain is $-40°F$ ($-40°C$) to $400°F$ ($205°C$). DOE 1090-2007 provides additional guidance for chain used at temperatures of less than $0°F$ ($-18°C$). At extreme temperatures, the load should be lifted slowly until it is slightly off the ground and the rigging inspected further before proceeding with the lift.

CHAIN SPECIFICATIONS

Nominal Chain Size*	Rated Load†		Material Diameter*	Length*	Width*	Approximate Weight† per 100'
	Grade 80	Grade 100				
7/32	2100	2700	0.217	0.67	0.32	54
9/32	3500	4300	0.275	0.88	0.41	73
5/16	4500	5700	0.330	1.02	0.48	100
3/8	7100	8800	0.397	1.22	0.57	148
1/2	12,000	15,000	0.520	1.56	0.75	250
5/8	18,100	22,600	0.630	1.93	0.92	377
3/4	28,300	35,300	0.787	2.23	1.07	570
7/8	34,200	42,700	0.875	2.25	1.14	730
1	47,700	—	1.000	3.07	1.49	985
1¼	72,300	—	1.260	3.92	1.74	1570

* in in.
† in lb

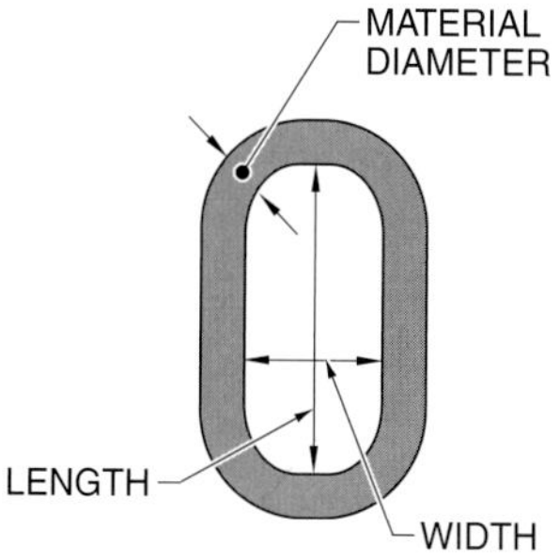

MATERIAL DIAMETER

LENGTH

WIDTH

Figure 4-16. Grade 80 and Grade 100 chain used for rigging applications are manufactured to standard specifications.

REDUCTION OF CHAIN RATED LOADS DUE TO TEMPERATURE

Chain Temperature*	While at Temperature		Permanently	
	Grade 80	Grade 100	Grade 80	Grade 100
Below −40	Consult manufacturer			
−40 to 400	None			
400	10%	15%	None	None
500	15%	25%	None	5%
600	20%	30%	5%	15%
700	30%	40%	10%	20%
800	40%	50%	15%	25%
900	50%	60%	20%	30%
1000	60%	70%	25%	35%
Over 1000	Remove from service			

* in °F

Figure 4-17. While chain slings can be used with a wider range of temperatures than any other type of sling, extreme temperatures do have a negative effect on rated loads.

Temperatures higher than 500°F (260°C) can cause a permanent reduction in the rated load of a chain and its sling fittings due to the annealing effect from the heat. Table 9-1.8.1-1 from ASME B30.9 or the recommendation of the sling's manufacturer should be used in determining the temporary and permanent rated load reductions of chain slings used under high-temperature conditions.

CHAIN INSPECTION

When performing an inspection, a chain must be cleaned of all dirt and grease, which may hide signs of damage. Each link should be inspected for cuts, gouges, and nicks. These can eventually lead to cracks that cause link fractures. **See Figure 4-18.** A *fracture* is a crack in metal caused by the stress and fatigue of repeated pulling or bending forces.

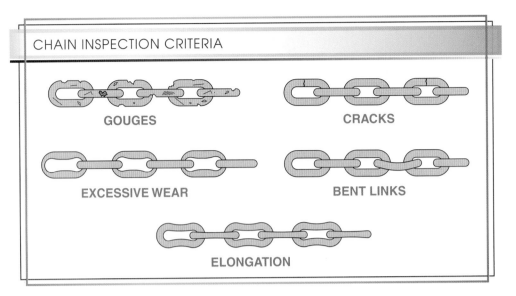

Figure 4-18. Damage and excessive wear to a chain or a chain sling's other components are reasons to remove them from service.

Overloading or sudden loading can deform a chain by elongating the link shapes, giving them a resemblance to figure eights. Lifting chain is designed to elongate before fracturing as much as 15% to 30%. Some amount of deformation is acceptable, but chain should be removed from service if the elongation exceeds 5%. This elongation is checked by measuring a length of chain and comparing the length to the chain specification. **See Figure 4-19.** If elongation is visually apparent, the chain is likely excessively deformed and must be replaced. Chain should also not be used if bent or elongated links prevent the chain from seating or flexing properly.

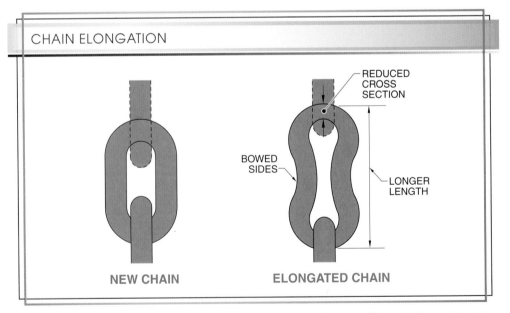

Figure 4-19. Repeated loading eventually causes a chain to elongate. Signs of elongated chain include reduced cross-section size of the link material, bowed sides, and longer link length.

Repeated loading, even at less than the rated load, also stretches link material, causing it to become thinner. Thinning at the links' contact points can also be caused by repeated flexing and wear. Link rod diameter should be measured and compared to the minimum allowable diameter for the chain size. According to ASME B30.9, chains should be removed from service if the thickness of any part of a link has decreased by 10% or more. **See Figure 4-20.**

CHAIN LINK WEAR

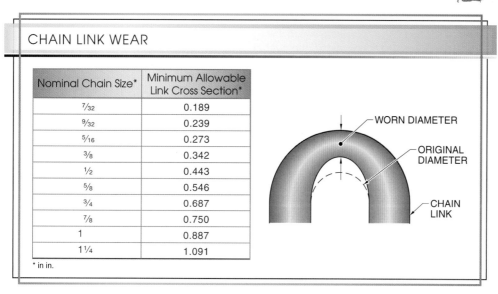

Nominal Chain Size*	Minimum Allowable Link Cross Section*
7/32	0.189
9/32	0.239
5/16	0.273
3/8	0.342
1/2	0.443
5/8	0.546
3/4	0.687
7/8	0.750
1	0.887
1 1/4	1.091

* in in.

Figure 4-20. Excessive chain wear causes the ends of links to become thinner. Worn chain should be checked for minimum allowable thickness.

When inspecting a chain, the links should be able to freely rotate from side to side approximately 90° without the links binding. Binding is an indication that links have stretched because the sides bow inward and prevent adjacent links from moving freely. A measurement of the length of a sling can be made by hanging it from a hook or placing it under slight tension and then measuring the distance from the inside of the upper end fitting to the inside of the lower end fitting. This dimension can be compared to the original length of the sling indicated on the identification tag.

Stretching of the chain from its original length is a sign of elongation, which indicates that the chain should be checked more thoroughly. Chain should be removed from service if any of the following defects are noted:
- missing or unreadable identification
- fractures
- deformed, bent, or twisted links
- gouges, nicks, or excessive wear
- elongated components or links
- evidence of heat damage or weld splatter
- excessive corrosion or pitting
- lack of free movement
- damaged hooks or hardware
- visible damage that causes doubt about continued use

COMMON APPLICATIONS— LIFTING TRANSFORMER VAULTS

A transformer vault is an underground precast concrete utility structure that is designed for the placement of transformers. After proper site planning, preparation, and excavation, a transformer vault is lifted into place with a crane or digger derrick. Typically a four-leg wire rope sling connected to a master link is used to lift a transformer vault into place. Each leg of the wire rope sling is secured to the transformer vault with a shackle and eyebolt. To install a transformer vault apply the following procedure:

1. Plan and prepare the site for installation.
2. Excavate the area where the transformer vault is to be installed.
3. Prepare the proper base and bedding.
4. Lift the base of the vault into place with a four-leg wire rope sling.
5. Level the transformer vault.
6. Install the transformers.
7. Install the gasket around the top edge of the vault.
8. Lift the top of the vault into place with a four-leg wire rope sling.
9. Install the grade ring riser.
10. Install the access frame and cover for future access of the vault.
11. Backfill the area around the transformer vault.

Tindall Corporation

Chapter 4 Learner Resources

For additional information, visit
qr.njatcdb.org
Item # 1649

Lineworker Rigging Practices

SLINGS AND HITCHES 5

OBJECTIVES
- List and describe the various materials and configurations used to construct slings.
- Differentiate between the rated load, breaking strength, and safety factor of a sling.
- List the types of information typically included on sling identification tags.
- Describe various types of sling hitches and their advantages and disadvantages.
- Calculate a sling load based on load weight and rigging arrangement.

Slings are the most common devices used to connect a load to the hook of a crane or hoist. Various types of slings may be used. A sling is chosen based on the specific requirements or conditions of a particular lifting operation. A sling may be constructed of wire rope, chain, metal mesh, fiber rope, or synthetic fibers. The expected force on a sling must be determined before selecting the size and type of sling to ensure that the sling is not overloaded. Various types of hitches can be used to attach a sling to a load. Some reduce the rated load of the slings. However, they have certain advantages, such as helping to maintain control of a load.

SLINGS

A *sling* is a flexible length of load-bearing material that is used to rig a load. They are used frequently in situations in which a load and a crane hook cannot be directly connected. Slings are an intermediate connection between a load and hooks that are too large to fit into the openings of an attachment point on the load.

More importantly, the use of slings with other rigging hardware provides a variety of possible rigging arrangements in order to design a safe and efficient lift. For example, flexible slings can often be looped around loads that have no attachment points. Furthermore, the variety of sling types and lengths available provides a number of options for dealing with different scenarios and load arrangements, including being able to adjust and position the point of a lift so that it remains above the load's center of gravity (CG).

Sling Materials

The most common types of sling materials used for rigging and hoisting are chain, synthetic webbing, synthetic roundslings, and wire rope. **See Figure 5-1.** Fiber rope slings and wire mesh slings are also available but are not as widely used. Synthetic web slings made from nylon are typically used for line work.

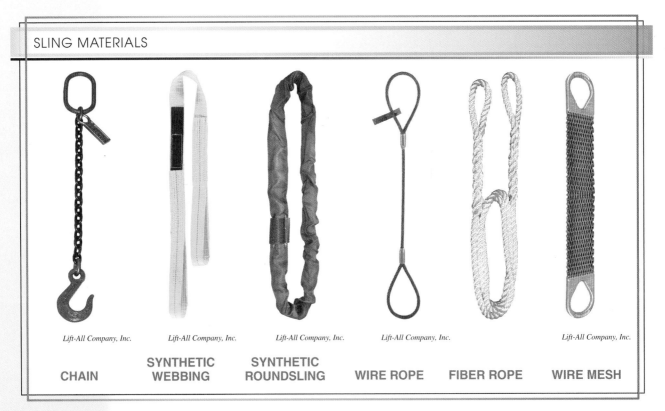

SLING MATERIALS

Lift-All Company, Inc.   *Lift-All Company, Inc.*   *Lift-All Company, Inc.*   *Lift-All Company, Inc.*   *Lift-All Company, Inc.*

| CHAIN | SYNTHETIC WEBBING | SYNTHETIC ROUNDSLING | WIRE ROPE | FIBER ROPE | WIRE MESH |

Figure 5-1. Slings are made from a variety of materials, each with advantages and disadvantages for different applications.

The type of sling selected for an application is based primarily on the strength and appropriateness of the material. Certain conditions may make some sling material unsuitable, perhaps even dangerous. For example, elevated temperatures or chemical exposure may weaken or damage certain sling materials. Sharp-edged or abrasive loads could damage or deform some sling types, particularly synthetic fiber slings.

Conversely, if the finished surface of a load might be damaged by a sling, a synthetic fiber sling may be recommended. Also, metal slings in certain arrangements are more likely to slip out of place under load, while synthetic fiber slings typically have greater friction with the load surface, which helps keep them in place. The sling's manufacturer or the standard ASME B30.9, *Slings,* should be consulted for guidance on appropriate applications.

Sling Configurations

Slings are formed as either eye-and-eye or endless. **See Figure 5-2.** An *eye-and-eye sling* is a sling in which loops are formed at each end of the sling body by doubling over the material and securing it by sewing, weaving, or using a compression fitting. Eye-and-eye slings are typically made from wire rope or synthetic webbing. An *endless sling,* also called a grommet sling, is a sling formed by attaching the ends of a sling body together to form a continuous loop. These types of slings can be made from chain or synthetic fiber materials.

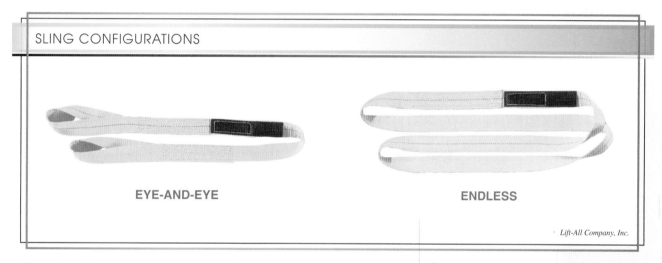

SLING CONFIGURATIONS

EYE-AND-EYE   ENDLESS

Lift-All Company, Inc.

Figure 5-2. Sling configurations are broadly categorized as either eye-and-eye or endless.

Other sling configurations include twisted eye, return (reverse) eye, triangle end, and transformer slings. A *twisted eye sling* is an eye-and-eye sling in which each eye forms a right angle to the plane of the sling body. Twisted eye slings are typically used for choker hitches. A *return eye sling,* also known as a reverse eye sling, is a sling that has multiple widths of webbing held edge to edge and in which each eye forms a right angle to the plane of the sling body. Return eye slings are typically used for basket and choker hitches. A *triangle end sling* is a sling with metal triangle-shaped fittings on each end. An *adjustable transformer sling* is a sling with adjustable eyes at each end to accommodate various types and sizes of transformers. Each end typically has a crown splice to prevent the eye from pulling apart.

Various types of end fittings are used with slings to make them more suitable for specific devices, such as thimbles, rings, or hooks. **See Figure 5-3.** These can be included on either eye-and-eye or endless slings. Some are permanently attached when the sling is manufactured, and some can be added by the rigger.

However, the size of a sling eye affects its suitability for use with other hardware. A sling eye must not be opened too widely, even if it can be fitted onto hardware because it strains the eye construction. **See Figure 5-4.** Synthetic web sling eyes must not be put around objects that are wider than one-third the length of the eye. For wire rope slings, the limit is one-half the length of the eye and not less than the diameter of the wire rope.

END FITTING EXAMPLES

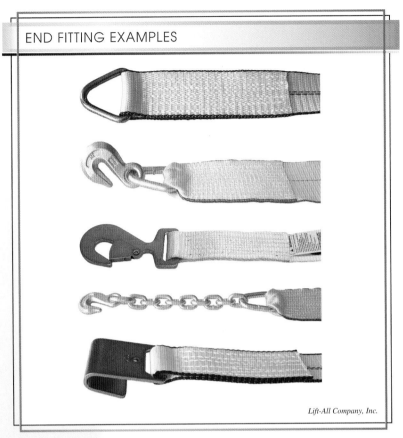

Lift-All Company, Inc.

Figure 5-3. A wide variety of end fittings are available for slings. Most types are D-shaped rings with various hooks.

Sling Rated Loads

Rigging components, such as hardware and slings, are rated by the amount of force they can safely tolerate. The *rated load* is the maximum tension that a rigging component may be subjected to while maintaining an appropriate margin of safety. This is also commonly called the load rating, rated capacity, or working load limit. The rated load of a component is different from its breaking strength.

Breaking strength is the tension at which a material is expected to break. The breaking strength of a particular material is determined by subjecting multiple samples to destructive load testing and averaging the results. However, the breaking strength value cannot be used directly as a safe value for lifting. There would be no margin of safety to compensate for slight manufacturing variations, underestimated load weight, sling age, or other weakening conditions. It is too likely that the forces will exceed the material's breaking strength, which can lead to failure.

A *safety factor,* also known as the design factor, is the ratio of a component's breaking strength to its rated load. **See Figure 5-5.** Chain slings use a minimum safety factor

LIMITS OF SLING EYE OPENINGS

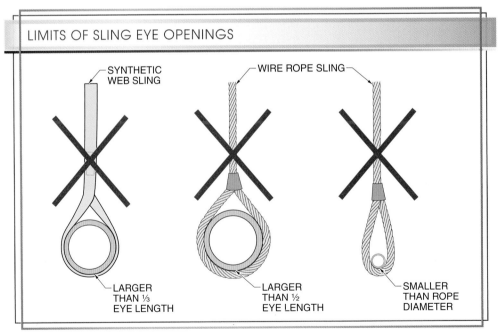

SYNTHETIC WEB SLING

WIRE ROPE SLING

LARGER THAN ⅓ EYE LENGTH

LARGER THAN ½ EYE LENGTH

SMALLER THAN ROPE DIAMETER

Figure 5-4. The size of a sling eye determines the range of component sizes that may be safely attached to it.

of 4, while other types of slings use a minimum safety factor of 5. Therefore, the breaking strength is four to five times greater than the rated load. For example, if a safety factor of 5 is used for a rigging assembly with a breaking strength of 5000 lb, the rated load of the rigging assembly is 1000 lb (5000 lb ÷ 5 = 1000 lb). Note that the rated loads of slings with permanently attached end fittings, such as hooks or rings, are for the entire sling assembly.

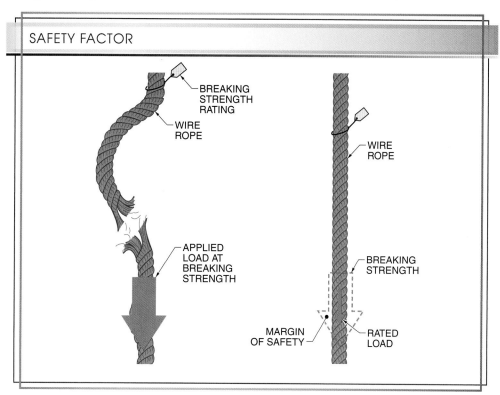

SAFETY FACTOR

Figure 5-5. The safety factor provides a large margin of safety between a component's breaking strength and the load weight it is rated to support.

Standard safety factors are only valid for steady loading, which requires slow and gradual hoisting and load travel. Dynamic loading involves rapid motions, such as quick changes in direction, load shifting, and load swinging, and should always be avoided. However, if the possibility cannot be avoided, such as when lifting in windy environments, the rigging equipment's rated load should be derated to approximately 60% of its specification in order to maintain a sufficient operating safety factor.

Sling Identification

Slings are tagged by their manufacturers to indicate relevant information about rated loads and other characteristics. **See Figure 5-6.** Regardless of the type of sling used, the identification tag must be legible and remain attached to the sling for its entire usable life. A sling should be removed from service if its tag is missing.

The amount of information on sling tags may vary. Some simply list the name or trademark of the manufacturer, sling material and length, and one rated load of the sling. However, other information may be included, often depending on the type of sling. Specific requirements for the information to be provided with each type of sling can be found in the standard ASME B30.9, *Slings*.

SLING IDENTIFICATION TAGS

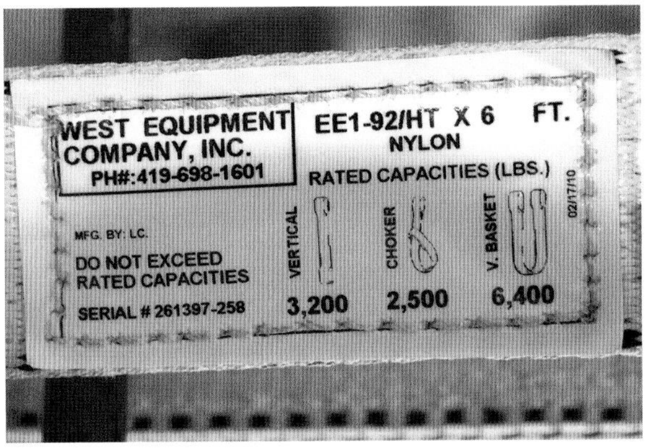

Figure 5-6. All slings must have legible identification tags that show basic information on the sling's manufacturer, type, and rated load.

Sling Application Rules

Correctly selecting and applying the appropriate slings and hardware are vitally important to ensure that loads are properly connected and safely hoisted. There have been numerous instances of equipment failure and subsequent death of personnel due to improperly rigged loads. To ensure safety, a qualified rigger must have a good working knowledge of the types of slings available and how to apply them under specific field conditions. Although the different types of slings have specific rules for their use, some general guidelines for using slings are the following:

- Slings must be selected to meet all of the requirements and conditions of a particular lift operation, including the rated load, the type of load, the type of hitch, the environment, and control of the load.
- If the proper type of sling or other rigging equipment is not available, the lift should be deferred until the proper equipment can be obtained.
- Synthetic fiber slings should never be used for hoisting personnel baskets.
- Body parts must never be placed between a sling and a load.
- Slings should never be shortened by knotting or twisting them.
- Slings should never be lengthened by tying them together.
- Loads should not be left resting on slings, which can damage the slings.
- If a sling is under a resting load, removal or repositioning should not be attempted until the load can be raised.
- Precautions should be taken to prevent the snagging or kinking of slings.
- Slings should not be dragged over abrasive surfaces.
- When removing a sling with a hoist, the commands should be initiated by the person who is extracting the sling.

There are many factors that can reduce the breaking strength of slings, including knots, used rope, sharp edges on loads, sharp bending around hooks and other attachments, and sling angles less than 30°.

SLING HITCHES

A *sling hitch* is an arrangement of one or more slings for connecting a load to a hoist hook. Most slings can be used in a variety of hitches, though this affects their effective rated load and appropriate applications. When deciding on a sling hitch for a load, consideration should be given to how the sling or slings will be attached and how control of the load will be maintained throughout the lift.

There are four basic types of sling hitches. **See Figure 5-7.** A *vertical hitch* is a sling hitch in which one end of a sling connects to a hoist hook and the other end connects to a load. The sling does not wrap around or under the load. A *basket hitch* is a sling hitch in which a sling is passed under a load, and both ends of the sling are connected to a hoist hook. A *choker hitch* is a sling hitch in which one end of a sling is wrapped under or around a load, passed through the eye at the other end of the sling, and then connected to a hoist hook. A *bridle hitch* is a sling hitch in which two or more slings share a common end fitting that is used as a lifting point.

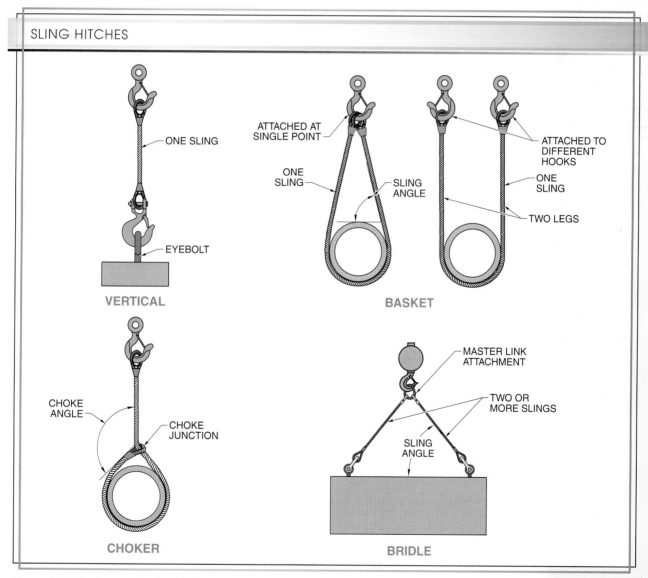

SLING HITCHES

Figure 5-7. The four basic sling hitches provide different methods for attaching a sling to a load.

The type of hitch used for a given application is determined by a number of factors. A load with no preexisting attachment points usually requires the use of a basket or choker hitch. Choker hitches are also used to bundle a number of pieces of material for a single lift. Basket hitches are commonly used when multiple hoist hooks are required.

Both choker and basket hitches may also involve additional wraps of the sling around the load in order to shorten the effective length of the sling or to provide additional friction between the sling and the load. Care should be taken that the wraps of the sling body do not cross and create a pinch point, which would reduce the strength of the sling. For web slings, the additional wraps of the sling may overlap but should not cross.

Vertical Hitches

The vertical hitch is the simplest rigging arrangement. One end of the sling is connected to a hoist hook and the other end to a load. **See Figure 5-8.** The vertical hitch allows the use of the full rated load of a sling (provided the attachments to the hook and the load are properly made) because there are no horizontal forces affecting the sling. However, a vertical hitch requires some type of attachment point on a load, such as a ring, an eyebolt, or a lug.

The use of a single vertical hitch is more likely than other hitches to result in an unstable load, particularly with wide loads, unless the attachment point is considerably higher than the load's CG. Any unbalance or shifting of a load can create an unsafe situation, unless a corrective force can be applied to keep the load level. For safety, control of the load must be established at the beginning of a lift and maintained throughout the lift. Therefore, a single vertical hitch configuration should only be used under a limited set of circumstances. For example, if the load will be hoisted not more than a few feet above the ground, a worker can usually provide the needed corrections to keep the load level. Otherwise, a tag line should be used.

A sling's rated load in a vertical hitch is typically the reference value that is derated when the sling is used at angles and in other arrangements. Identification tags cannot practically include rated load information for every possible hitch and angle but will always include the vertical hitch rated load. With this information, rated loads for other hitches can be determined.

Basket Hitches

A basket hitch is formed by passing a sling underneath a load and then connecting both ends of the sling to a hook, or two hooks. Basket hitches can be used to cradle a load for lifting. **See Figure 5-9.** An advantage of a basket hitch is that it does not rely on lift point hardware on a load, such as eyebolts or lifting eyes. However, it does require

VERTICAL HITCHES

Figure 5-8. Vertical hitches connect a load to a lifting hook at a single lift point. Rigging assemblies may include multiple vertical hitches.

space under a load both before and after a lift so the sling can be passed under the load.

Another advantage of a basket hitch is that it effectively doubles the rated load of the sling as long as the legs of the sling remain within 5° of vertical. This is because the load weight is divided between the two halves of the sling. However, if the ends of the sling are not vertical, a reduction of the rated load must be made based on the angle of the legs of the sling.

When grouping sling eyes together, such as in basket and bridle hitches, special safety considerations apply. **See Figure 5-10.** First, they should always be connected to a common hook or fitting, not to each other. For example, forming a basket hitch by connecting one sling eye to the other eye induces side loads on the second eye, which can cause it to fail. Second, the maximum allowable angle between the slings in a hook is 90°. If the angle formed by the slings is greater than 90°, a shackle or master link must be used as an interfacing device between the slings and the hook. Shackles and master links can accommodate sling angles up to 120°. A shackle used in this application should be installed with its pin in the hook and the slings gathered in its bow.

BASKET HITCHES

Lift-All Company, Inc.

Figure 5-9. A basket hitch requires one end of a sling to be passed under a load and back up to a lifting hook.

Most basket hitch rigging requires two or more baskets. A single basket hitch is only appropriate with a load shape that allows the basket to remain completely above the load's CG. Otherwise, the risk of slipping is too great.

A factor to consider when using wire rope slings in a basket or choker hitch configuration is the size of the bend used to wrap around the load. A very small bend can put extra stress on the wire rope or even cause a permanent kink. The wire rope specifications should be referred to for information on bending efficiency.

Choker Hitches

Choker hitches are commonly used to attach slings to bundles of long materials, such as beams, tubing, and lumber. **See Figure 5-11.** When tension is applied, the sling constricts around the load, tightening the bundle and making it easier to manage.

However, the trade-off is that the bend created in the sling when forming the hitch reduces the sling's rated load. The amount of reduction depends on the choke angle. **See Figure 5-12.** The *choke angle* is the angle formed at the choke between the vertical part of a sling and the part of the sling surrounding a load. Most slings used in a choker hitch retain 100% of their rated load if the choke angle is greater than 120°.

For smaller choke angles, the rated load is reduced to a percentage of the vertical or bridle rated load. As the choke angle gets smaller, the bend in the sling is more severe, and the rated load is reduced further. These percentages are applicable for synthetic fiber and wire rope slings. For chain and other sling types, the manufacturer must be consulted. Note that if a choker sling is used at an angle, this derating applies to the rated load for that angle.

FORMING PROPER BASKET HITCHES

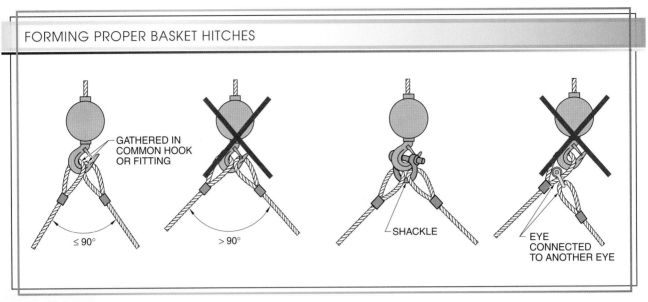

GATHERED IN
COMMON HOOK
OR FITTING

≤ 90°

> 90°

SHACKLE

EYE
CONNECTED
TO ANOTHER EYE

Figure 5-10. When forming a basket hitch, sling eyes must be grouped in a proper manner to prevent undue stress on the slings and slippage from the hook.

CHOKER HITCHES

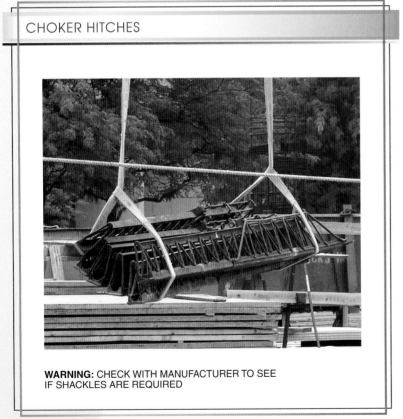

WARNING: CHECK WITH MANUFACTURER TO SEE
IF SHACKLES ARE REQUIRED

Figure 5-11. Choker hitches are particularly good for lifting bundles of long materials. The hitch tightly binds the materials to secure the load for lifting.

When applying a choker hitch, the eye of the sling that forms the choker should never be beaten down against the load. This is sometimes done to tighten the hitch and increase friction between the sling and the load. However, this action decreases the choke angle, which reduces the rated load of the sling.

A better practice is to use a longer sling and wrap it completely around the load one or more times. Also, a block or other support material may be secured to the top of the load for the lower eye of the sling to bear against and maintain the proper choke angle. This supporting device should be secured so that it does not slip if tension on the sling is released during the hoisting operation. Finally, the sling should be long enough so that the choke is on the body of the sling and not on sling fittings.

Rigging a load with a single choker hitch should be avoided because the load can shift and easily slip out of the hitch. If the single choker hitch configuration is necessary, a synthetic fiber sling with multiple wraps can be used to increase the friction between the sling and the load.

CHOKER HITCH DERATING*

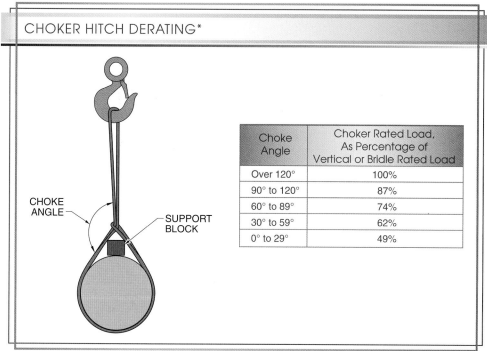

Choke Angle	Choker Rated Load, As Percentage of Vertical or Bridle Rated Load
Over 120°	100%
90° to 120°	87%
60° to 89°	74%
30° to 59°	62%
0° to 29°	49%

* for synthetic and wire rope slings; for chain and other slings, consult manufacturer

Figure 5-12. Choker hitches put extra stress on a sling at the point of choke, which reduces the sling's rated load.

While most loads are hoisted in a horizontal orientation, some require hoisting in a vertical orientation, which presents additional difficulties. Approved below-the-hook devices, such as clamps, can be temporarily added to the load as mechanical stops. These stop the sling from shifting position or sliding off a vertical load.

Choker hitches can cause significant wear on the inside of the eye where it contacts the sling body, particularly for wire rope slings. To prevent this damage, a shackle should be used in the lower eye to form the choke. **See Figure 5-13.** A shackle would resist wear longer than the sling eye and is easier to replace when necessary. A shackle should be connected with the pin inside the sling eye so that pulling on the body of the sling does not cause the pin to unscrew.

SHACKLES IN CHOKER HITCHES

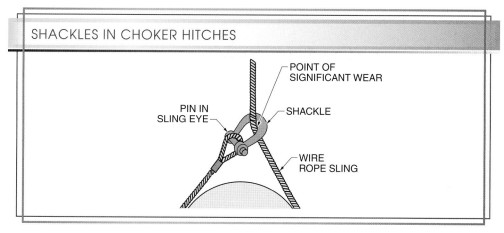

Figure 5-13. Shackles should be used to form choker hitches so that a sling eye does not wear prematurely.

COMMON APPLICATIONS— INSTALLING PAD-MOUNTED TRANSFORMERS

A pad-mounted transformer is encased in a steel enclosure mounted on a concrete, fiberglass, or polymer pad that is placed on the ground. Pad-mounted transformers are designed to be tamperproof, are typically secured with a padlock, and must have a clearly visible warning sign indicating high-voltage electrical equipment. Pad-mounted transformers are used with underground power distribution systems and are typically installed in public areas and residential easements.

To install a pad-mounted transformer, apply the following procedure:
1. Prepare the area.
2. Attach the proper lifting attachments and sling to the transformer pad.
3. Place the transformer pad into position.
4. Attach a sling to the preinstalled lifting points on the transformer.
5. Place the transformer into position on the transformer pad.

Bridle Hitches

A bridle hitch allows a load to be secured at more than one point, which increases load stability and allows greater control of the load. **See Figure 5-14.** Most bridle hitches consist of a group of slings connected to attachment points on a load similar to vertical hitches, except at an angle. The tops of the slings are gathered at a common hook or fitting. However, bridle hitches can also consist of a group of choker hitches and basket hitches. They become bridle hitches because they are similarly joined at the top.

A bridle hitch is rigged with a bridle sling. A *bridle sling* is an assembly of two or more sling legs, each with an end gathered together at a common end fitting. Bridle slings may be purchased as a prefabricated assembly or assembled in the field from individual slings.

Prefabricated Bridle Slings. Bridle sling assemblies are often manufactured as a unit that cannot be disassembled or modified in the field. **See Figure 5-15.** These assemblies have sling legs joined at a master link, which is a steel ring. The master link must be rated to carry the combined load of all of the sling legs. If a single rated load is provided for a bridle sling, it is assumed to be for a 60° bridle hitch.

If a prefabricated bridle sling has more legs than needed for an application, the unused legs should be secured to prevent them from interfering with the load or other rigging during the hoisting operation. The loose ends can be connected to the master link with a shackle.

BRIDLE HITCHES

Figure 5-14. A bridle hitch is formed from multiple slings that are attached to the load at one end and gathered together at the other end, forming a triangle. Some rigging assemblies include multiple bridle hitches.

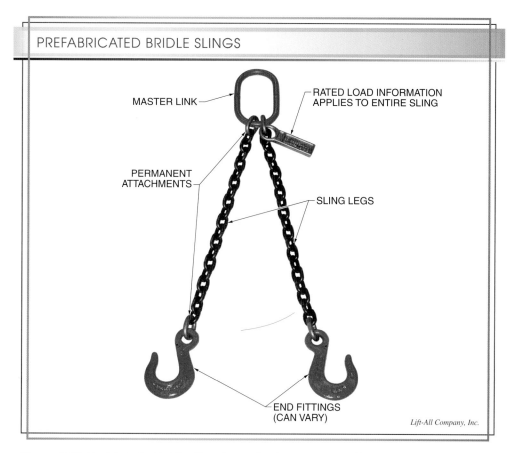

PREFABRICATED BRIDLE SLINGS

MASTER LINK

RATED LOAD INFORMATION
APPLIES TO ENTIRE SLING

PERMANENT
ATTACHMENTS

SLING LEGS

END FITTINGS
(CAN VARY)

Lift-All Company, Inc.

Figure 5-15. Prefabricated bridle slings have sling legs permanently attached to a master link.

Field-Assembled Bridle Slings. Bridle hitch configurations can also be assembled from individual slings and rigging hardware. When slings are grouped in a hook, they should sit in the base or bowl of the hook, not on the tip, and not bind or overlap each other. Synthetic fiber slings are particularly prone to pinching or bunching due to limited space on a hook. To avoid these problems, a larger shackle or master link should be used. **See Figure 5-16.**

If a bridle hitch incorporates basket or choker hitches, special considerations apply. First is the potential slippage of slings toward the center of the load. Second, when using synthetic web slings in low sling angles, the load may bear on only one edge of the sling material, which can damage the sling. **See Figure 5-17.** Both problems can be alleviated by using longer slings in order to increase the attachment sling angle. Alternatively, spreader or lifting beams can be used to avoid these problems altogether.

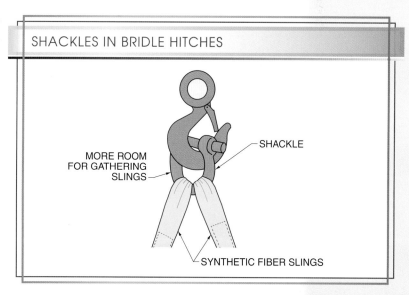

SHACKLES IN BRIDLE HITCHES

SHACKLE

MORE ROOM
FOR GATHERING
SLINGS

SYNTHETIC FIBER SLINGS

Figure 5-16. Shackles can be used to form bridle slings in the field, particularly when all the sling eyes will not safely fit within the hoist hook.

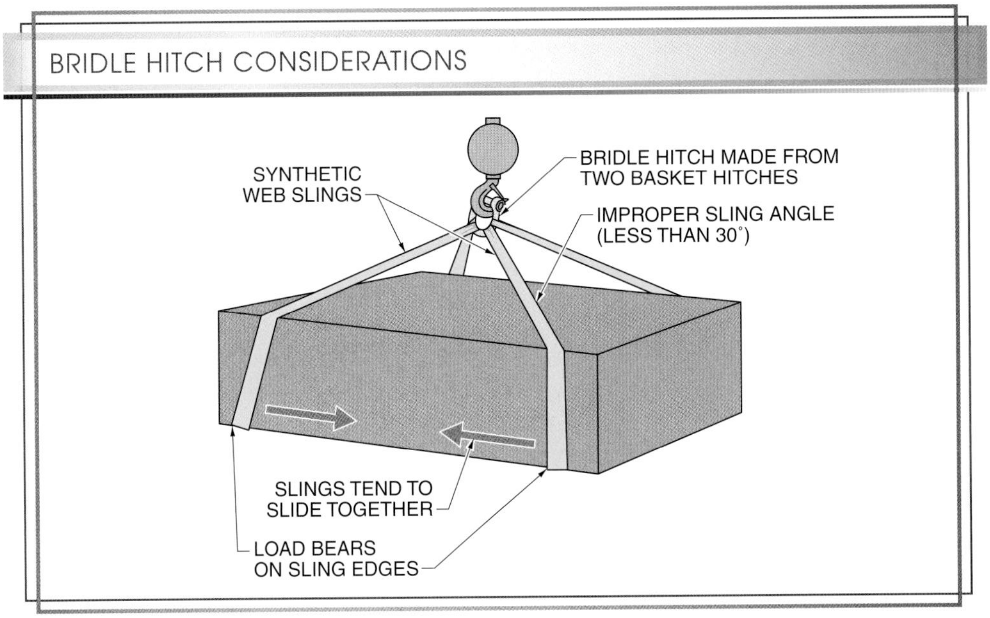

BRIDLE HITCH CONSIDERATIONS

SYNTHETIC
WEB SLINGS

BRIDLE HITCH MADE FROM
TWO BASKET HITCHES

IMPROPER SLING ANGLE
(LESS THAN 30°)

SLINGS TEND TO
SLIDE TOGETHER

LOAD BEARS
ON SLING EDGES

Figure 5-17. When using a pair of basket hitches as a bridle hitch, they can easily slide together under load, possibly causing load instability. Also, if synthetic web slings are used, the load may bear disproportionately on one edge, causing sling damage.

SLING LOADS

When selecting slings for a particular rigging application, the primary consideration is their rated loads. Each sling must be rated to withstand, with the appropriate safety factor, the maximum force that it will be subjected to in the proposed rigging configuration. Determining this sling loading involves two major factors: the portion of load weight it must support and the sling angle.

A vertical sling experiences only the force of the portion of load weight it is supporting. **See Figure 5-18.** However, a sling loaded at an angle also experiences horizontal forces. The horizontal force is proportional to the sling angle. A smaller angle induces greater force. The vertical and horizontal forces combine to result in a total load on the sling that can be much greater than its portion of the load weight.

The sling angle is used to adjust sling rated loads for nonvertical hitch configurations. The *sling angle,* also known as the angle of loading or horizontal angle, is the acute angle between horizontal and the sling leg. The minimum recommended sling angle is 30°. Even if a sling had the rated load to withstand the extreme horizontal forces at angles less than 30°, the load or the lift points may still be damaged. If a sling angle of less than 30° is unavoidable, additional calculations must be made by a qualified person to determine whether these forces can be sustained by the load. Alternatively, a lifting beam can usually be used in order to avoid compressive forces on the load.

Based on the sling angle and the portion of the load weight supported, the actual tension on the sling can be calculated. There are a number of methods for doing this.

Load Triangles

The most precise way to calculate the tension on a sling involves a load triangle. A *load triangle* is the right triangle formed by an angled sling and the horizontal and vertical forces acting on the sling. **See Figure 5-19.** The tension force along the sling can be calculated using simple geometry rules.

SLING LOADS

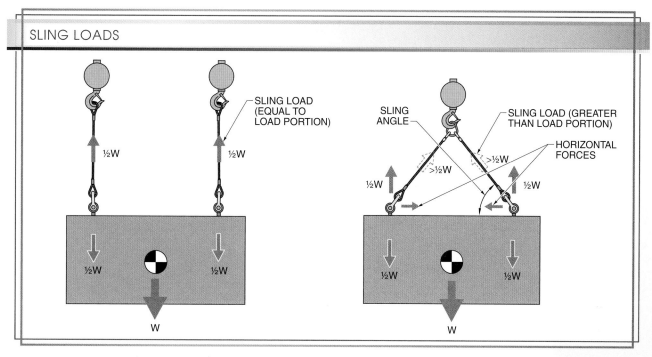

Figure 5-18. Vertical hitches experience tension equal to the weight they support. Slings used at an angle, however, experience tension greater than the load weight due to the additional horizontal forces.

LOAD TRIANGLES

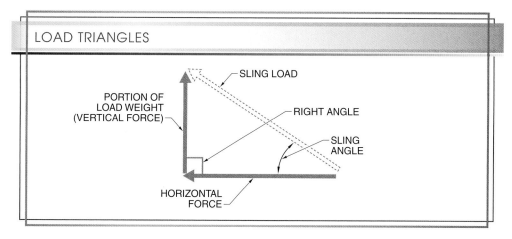

Figure 5-19. Determining a sling load requires consideration of the portion of the load weight supported and the horizontal force. Together, these forces form a load triangle.

First, the portion of the load weight supported by the sling must be determined. This is the vertical force component. This vertical force is then multiplied by a multiplier in order to determine the sling tension. This multiplier is based on the sling angle, so it changes for different sling geometries. Greater sling angles produce smaller multipliers and less sling tension, while smaller sling angles produce larger multipliers and greater sling tension. The multiplier can be determined by two different methods.

L/H Ratio. The sling tension is proportional to the load weight portion in the same way that the sling length is proportional to the vertical distance between the load and the point of sling attachment. **See Figure 5-20.** When rearranged, the ratio between the sling length and rigging height becomes a simple multiplier for converting the portion of the load weight to sling tension. This L/H ratio is shown in the following formulas:

$$\frac{F_{sling}}{W_{portion}} = \frac{L}{H}, \text{which is rearranged as } F_{sling} = \frac{L}{H}W_{portion}$$

where

F_{sling} = tension force on sling (in lb or t)

L = length of sling (in ft or in.)

H = vertical distance between lift point and upper point of attachment (in ft or in.)

$W_{portion}$ = portion of load weight supported by sling (in lb or t)

L/H RATIOS

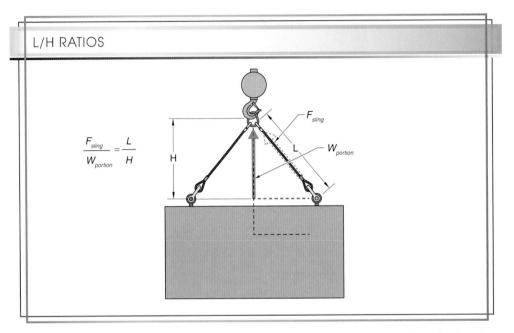

Figure 5-20. The L/H ratio can be used to easily calculate a sling load based on the lengths of the sides of the load triangle.

For example, a symmetrical 10,000 lb load is to be rigged with two 8′-long slings. The point of attachment is 6′ above the load. Since it is symmetrical, the load supported by each sling will be 5000 lb.

$$F_{sling} = \frac{L}{H}W_{portion}$$

$$F_{sling} = \frac{8}{6}5000$$

$$F_{sling} = 1.33 \times 5000$$

$$F_{sling} = \textbf{6650 lb}$$

Therefore, the slings and any other hardware used for hoisting this load must have a rated load of at least 6650 lb each in order to lift the load safely.

The use of shorter slings increases the L/H ratio because it significantly decreases the vertical distance. For example, if 6′ slings are used with the same attachment points, the L/H ratio is approximately 2.12. Therefore, the load on each sling would be 10,600 lb (2.12 × 5000 = 10,600). Note that an L/H ratio greater than 2 indicates that the sling angle is less than 30°, which is not recommended.

This method can be used even if the actual length of the slings and the height of attachment are unknown. A ruler or other scale can be used from a distance, and the relative length of the sling can be compared to the height of the attachment. For example, a ruler is held at arm's length and aligned with the sling. The apparent length of the sling is 6″. The ruler is then repositioned to determine the relative height of the attachment, which is observed to be 5″. The L/H ratio is therefore 6/5 or 1.2. If the load weighs 6000 lb, then the tension on each sling is 3600 lb (1.2 × 3000 = 3600).

Sling Angle Multiplier. Alternatively, the sling load multiplier is equal to the cosecant of the sling angle. *Cosecant* is a trigonometric function equal to the ratio between the hypotenuse and the opposite legs of a right triangle. The resulting sling tension calculation is represented in the following formula:

$$F_{sling} = \csc \alpha \times W_{portion}$$

where

F_{sling} = tension force on sling (in lb or t)

α = sling angle (in °)

$W_{portion}$ = portion of load weight supported by sling (in lb or t)

In the case of a load triangle, the cosecant is the same as the L/H ratio, which means that the result is the same multiplier as calculated with the L/H method. However, sometimes the sling angle multiplier method is easier to use. The sling angle can be measured accurately using a protractor level, and the cosecant can be easily calculated with a scientific calculator or looked up in a trigonometric values chart. **See Figure 5-21.** The cosecant function is the reciprocal of the sine function. If a calculator lacks the cosecant [csc] function, the sine [sin] and then reciprocal functions can be used instead.

SLING ANGLE MULTIPLIERS

Angle	Sine	Cosecant
90°	1.000	1.000
85°	0.996	1.004
80°	0.985	1.015
75°	0.966	1.035
70°	0.940	1.064
65°	0.906	1.103
60°	0.866	1.155
55°	0.819	1.221
50°	0.766	1.305
45°	0.707	1.414
40°	0.643	1.556
35°	0.574	1.743
30°	0.500	2.000

Figure 5-21. A sling angle multiplier can be used to easily calculate a sling load based on the sling angle.

For example, a protractor level is used to determine that a sling angle is 37°. The cosecant of the angle is 1.662. If the load weighs 12,000 lb, and each sling supports half of the load, what is the tension on each sling?

$$F_{sling} = \csc \alpha \times W_{portion}$$
$$F_{sling} = \csc 37° \times 6000$$
$$F_{sling} = 1.662 \times 6000$$
$$F_{sling} = \mathbf{9972\ lb}$$

Sling Load Charts

Sling load calculations can be used to determine actual sling tension, which is compared to a sling's vertical rated load to determine if the sling is appropriate for a particular rigging arrangement. Alternatively, the rated load for a sling in a particular rigging arrangement can be looked up in a sling load chart. Note, however, that these charts may be used only for symmetrical rigging arrangements.

A sling load chart includes the rated loads for various sizes of a certain type of sling in different hitches, including angled slings in bridle hitches. **See Figure 5-22.** Multiple sling angles are shown, usually the standard sling angles of 90°, 60°, 45°, and 30°. This results in an approximated rated load that provides extra safety factor. For rigging using angles between those shown on the chart, the smaller angle in the chart is used.

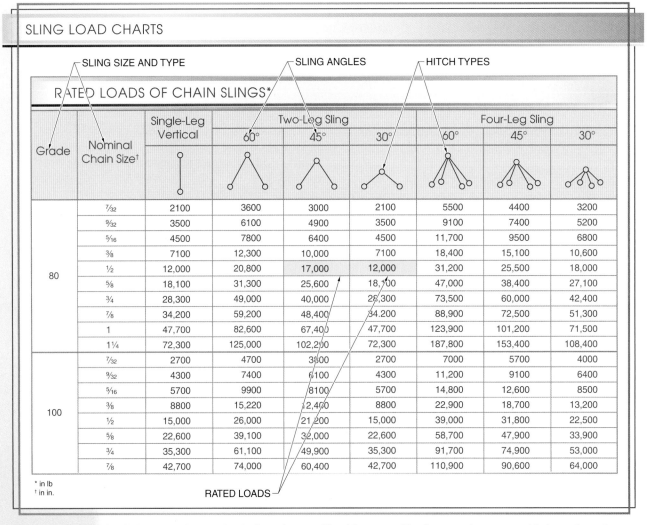

Figure 5-22. Sling load charts can be used to look up the rated load for a specific sling type in common hitch configurations.

For example, a load is rigged, and the sling angle is 52°. Since the load chart does not include this angle, the next lower angle of 45° is used. If a two-leg, ½″, Grade 80 alloy chain sling is used, the 45° rated load is 17,000 lb. Therefore, the load could weigh up to 17,000 lb and be hoisted safely using this sling assembly. If the sling angle was 38°, the rated loads for a sling angle of 30° would be used, and the safe working load for the sling assembly would be only 12,000 lb.

Sling load charts are included in the standard ASME B30.9, *Slings,* for all types of slings. Selected sling load charts are also often reprinted by manufacturers and safety agencies for job site use. However, note that some manufacturers publish sling load charts that apply to their products only. In particular, color code charts for synthetic roundslings are not universally applicable. Therefore, slings must be carefully checked to ensure that they are the same type as those referenced by the charts.

Lineworker Rigging Practices

SLING TYPES

6

OBJECTIVES

- List the advantages and disadvantages of slings made from synthetic fibers.
- Describe the construction of web sling materials and how they are fabricated into lifting slings.
- Identify the characteristics of web slings that affect the strength of the sling.
- Describe the construction of round-slings and how they are fabricated in to lifting slings.
- Identify the inspection criteria for synthetic slings.
- Differentiate between various chain sling configurations.
- List the identification requirements for chain and chain slings.
- Describe the general procedure for seizing and cutting a wire rope.
- Compare the installation procedures and efficiency ratings of eye and socket terminations.

Slings are typically made from synthetic fiber, chain, or wire rope. Synthetic slings are among the most versatile of slings. They are available with highly rated loads but remain lightweight. They are also available in a variety of different sizes and configurations. Their material is soft and flexible, making them ideal for use with easily damaged loads. Chain slings are manufactured from lengths of chain with common rigging hardware components, such as hooks, permanently attached to the ends. Many chain slings are fabricated in multiple-leg configurations. Wire rope slings are strong and durable. They are not as flexible as synthetic slings but still adequate for use with most types of loads. Wire rope slings are available prefabricated with a variety of terminations and end fittings.

SYNTHETIC SLINGS

Synthetic slings are constructed from yarns of artificial fibers, typically polymer fibers. **See Figure 6-1.** These materials are used for their high strength-to-weight ratios. Plastics are the most familiar types of polymers. Rigging slings are typically made with either nylon or polyester yarns, though specialty slings may be made with other polymers.

SYNTHETIC SLINGS

Lift-All Company, Inc.

Figure 6-1. Synthetic slings are fabricated from artificial fibers, such as nylon or polyester.

Synthetic slings are soft and flexible, making them ideal for rigging loads with fine surface finishes or that are easily damaged. For example, they are often used with glass and porcelain insulators or loads with polished or painted surfaces. However, they also make good general-purpose slings. They are light, easy to store, and versatile. The primary disadvantage of synthetic slings is their susceptibility to abrasion and puncturing.

Synthetic fibers are generally resistant to most common industrial chemicals, which makes synthetic slings ideal for use in industrial environments. **See Figure 6-2.** Polyester is resistant to many acids, and nylon is resistant to most alkalis. Both are resistant to oils, detergents, water, and solvents. However, certain chemicals and environmental conditions may still cause degradation. Also, any metal end fittings likely have different chemical resistances than synthetic fibers. It is critical to consult the specifications or the manufacturer when selecting a sling to be used in chemically active areas.

Extreme temperatures are the primary environmental concern. Synthetic slings should not be exposed to temperatures above 200°F (93°C) or below −40°F (−40°C). This is a much narrower range than the acceptable temperature range for chain or wire rope. Also, exposure to ultraviolet light, such as from sunlight or arc welding, may weaken synthetic material without any visible indication.

GENERAL CHEMICAL RESISTANCES OF SYNTHETIC SLING MATERIALS		
Chemical	Polyester	Nylon
Acids	Most	No
Alcohols	Yes	Yes
Aldehydes	No	Yes
Alkalis, strong	Some	Yes
Alkalis, weak	Yes	Yes
Bleaching agents	Yes	No
Dry cleaning solvents	Yes	Yes
Ethers	Yes	Yes
Hydrocarbons	Yes	Yes
Ketones	Yes	Yes
Oils	Yes	Yes
Soap/detergents	Yes	Yes
Water/seawater	Yes	Yes

Figure 6-2. An advantage of synthetic slings is that the fibers are resistant to many types of chemicals.

Synthetic sling construction begins with synthetic fibers. Fibers are twisted into yarns and the yarns are twisted or woven into sling material. The fibers are also made into threads used for stitching parts of the slings together. The two most common types of synthetic slings are web slings and roundslings. Both are made from similar yarn materials but fabricated differently.

Web Slings

A *web sling* is a flat rigging sling made from synthetic webbing material. *Webbing* is flat, narrow strapping woven from yarns of strong synthetic fibers. **See Figure 6-3.** Webbing material is very strong and distributes pressure across a wide surface.

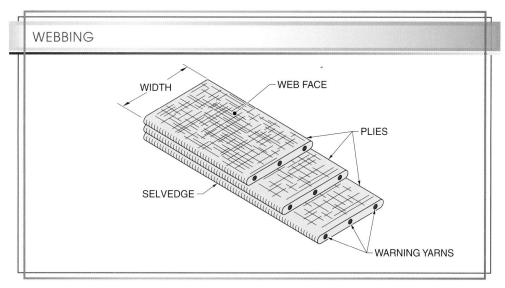

WEBBING

WIDTH

WEB FACE

PLIES

SELVEDGE

WARNING YARNS

Figure 6-3. Webbing material consists of yarns of synthetic fibers that are woven into wide, flat straps. Multiple layers, or plies, can be sewn together to make stronger webbing.

Webbing. Webbing for rigging purposes is made of woven nylon or polyester yarns. Colored marker yarns woven into the center of the face of the webbing may be used to identify the material. Nylon webbing has black markers or no markers. Polyester webbing has blue markers. Manufacturers may also use colored marker yarns at the edges of the webbing to indicate other materials or construction types.

Most web sling damage starts on the edge, so webbing includes selvedges for strength. A *selvedge* is an edge treatment on woven material that helps prevent unraveling. A selvedge is formed from continuous yarns that run back and forth across the width of the webbing, exposing only a turn of the yarn at the edges. **See Figure 6-4.** The weave is then drawn tighter at the edge to strengthen the material.

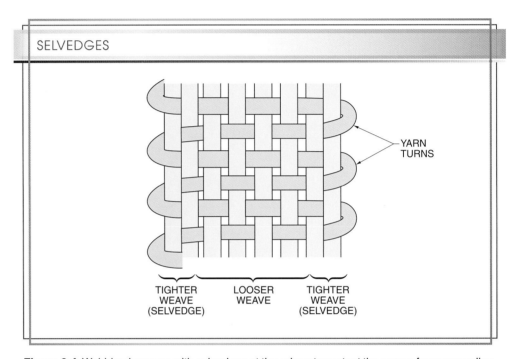

SELVEDGES

YARN TURNS

TIGHTER WEAVE (SELVEDGE) LOOSER WEAVE TIGHTER WEAVE (SELVEDGE)

Figure 6-4. Webbing is woven with selvedges at the edges to protect the weave from unraveling.

In some cases, special brightly colored yarns are woven into the webbing core as a safety feature. If the webbing is not damaged, these yarns are not visible. However, if webbing material is worn or torn to expose these yarns, the wear is considered excessive, and the webbing is no longer safe to use.

Webbing is available in multiple widths and numbers of plies (layers), which determine its strength. The wider and the greater number of plies in the webbing, the stronger it is. Rigging webbing ranges from 1″ to 12″ wide and is constructed in one, two, or four plies. The plies are stitched together in order to keep them aligned.

Webbing material is broadly classified by its breaking strength as Class 5 or Class 7 webbing. Breaking strength is determined by the thickness of the material, the type of material, and its construction. Class 5 webbing has a minimum certified tensile strength of 6800 lb/in. of width per ply. Class 7 webbing has a minimum certified tensile strength of 9800 lb/in. of width per ply. Either class of webbing may be manufactured into single- or multiple-ply webbing.

Web Sling Fabrication. Webbing material by itself cannot be used for rigging, as there is no safe way to attach it to other rigging components. Instead, a length of webbing is fabricated into a web sling by sewing its ends into a certain configuration. The ends are folded over and sewn into either large loops or small loops with metal end fittings. The stitching thread must be made from the same material as the webbing. The fittings should have smooth surfaces with no sharp edges that could damage the webbing.

A web sling is composed of several parts: the length, body, splices, and loop eyes. **See Figure 6-5.** The length is the distance between the ends of a web sling, including any fittings. The body is the portion of the sling that is between the loop eyes or any end fittings. A *splice* is an overlap of webbing material that is sewn together. A *loop eye* is a length of webbing folded back and spliced to the sling body, forming a closed loop. The length of a loop eye varies with the width of the sling webbing. When full-width sling loop eyes are too wide to properly fit into a hoist hook, a web sling with tapered loop eyes can be used. Web slings with tapered loop eyes have webbing folded to narrower widths at the bearing points to accommodate narrower hooks.

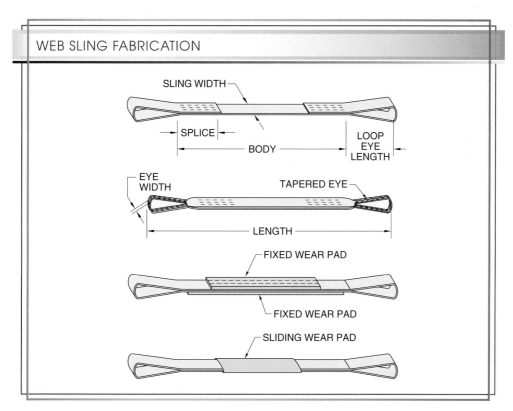

WEB SLING FABRICATION

Figure 6-5. Webbing material is made into web slings by splicing or forming loop eyes.

Wear pads are another common component of web slings. A *wear pad* is a leather or webbed pad used to cover the body of a web sling in order to protect it from damage. Wear pads are either sewn onto the webbing or slide along the body, allowing adjustable protection.

Web slings are fabricated in six standard configuration types. **See Figure 6-6.** Manufacturers may offer additional configurations for specialized applications. The six standard configuration types are as follows:

- Type I, also known as a triangle-choker (TC) sling, is a web sling made with a triangle fitting on one end and a large slotted choker fitting on the other end. Type I web slings are typically used for choker hitches, though they are also suitable for vertical and basket hitches.

WEB SLING CONFIGURATION TYPES

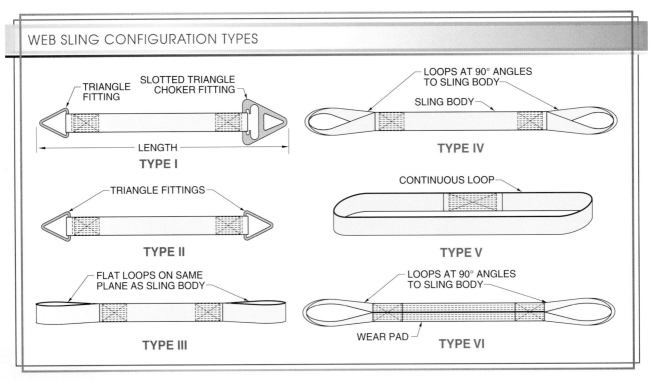

Figure 6-6. There are six primary web sling configurations, which are available in various sizes and webbing types.

- Type II, also known as a triangle-triangle (TT) sling, is a web sling made with a triangle fitting on both ends. Type II web slings are used for vertical or basket hitches.
- Type III, also known as an eye-eye (EE) sling, is a web sling made with large flat loops at each end that are on the same plane as the body. Type III web slings are used for vertical or basket hitches. They are also used for choker hitches by passing one eye around the load and through the other eye. Type III web slings are often available with tapered loop eyes to permit their use with hooks.
- Type IV is a web sling made with both loop eyes twisted to form eyes that are at right angles to the plane of the sling body. Because one loop eye does not need to be twisted in order to line up with the other end, this arrangement makes a slightly better choker hitch than the Type III style. Type IV slings can also be used for vertical and basket hitches.
- Type V, also known as a grommet sling, is an endless web sling made by joining the ends with a load-bearing splice. Type V web slings are used for numerous applications and are the most widely used. They may be used in basket, vertical, or choker hitch applications.
- Type VI, also known as a reverse eye (RE) sling, is an endless web sling with butted edges sewn together to form a body and two loop eyes, which are at right angles to the plane of the body. This sling is known as an RE sling because the eyes are formed by folding the webbing in the opposite direction than that used to form the loop eyes in a Type III or Type IV web sling. A wear pad is often added on either one side or both sides of the sling body. Type VI web slings are used for rugged service, such as lifting irregularly shaped objects like stones.

Web Sling Strength. Web sling specification tables show the rated loads for various webbing classes, webbing widths, number of plies, and hitch types. **See Figure 6-7.** These values are calculated using both fabrication efficiencies and safety factors.

RATED LOADS FOR WEB SLINGS

Class	Plies	Width†	Vertical*	Choker*	Bridle or Basket* Vertical*	60°	40°	30°	Type V Endless*
5	1	1	1100	880	2200	1900	1600	1100	2200
		1½	1600	1280	3200	2800	2300	1600	3200
		1¾	1900	1520	3800	3300	2700	1900	3800
		2	2200	1760	4400	3800	3100	2200	4400
		3	3300	2640	6600	5700	4700	3300	6600
		4	4400	3520	8800	7600	6200	4400	8800
		5	5500	4400	11,000	9500	7800	5500	11,000
		6	6600	5280	13,200	11,400	9300	6600	13,200
	2	1	2200	1760	4400	3800	3100	2200	4400
		1½	3300	2640	6600	5700	4700	3300	6600
		1¾	3800	3040	7600	6600	5400	3800	7600
		2	4400	3520	8800	7600	6200	4400	8800
		3	6600	5280	13,200	11,400	9300	6600	13,200
		4	8200	6560	16,400	14,200	11,600	8200	16,400
		5	10,200	8160	20,400	17,700	14,400	10,200	20,400
		6	12,300	9840	24,600	21,300	17,400	12,300	24,600
7	1	1	1600	1280	3200	2800	2300	1600	3200
		1½	2300	1840	4600	4000	3300	2300	4600
		1¾	2700	2160	5400	4700	3800	2700	5400
		2	3100	2480	6200	5400	4400	3100	6200
		3	4700	3760	9400	8100	6600	4700	9400
		4	6200	4960	12,400	10,700	8800	6200	12,400
		5	7800	6240	15,600	13,500	11,000	7800	15,600
		6	9300	7440	18,600	16,100	13,200	9300	18,600
		8	11,800	9440	23,600	20,400	16,700	11,800	23,600
		10	14,700	11,760	29,400	25,500	20,800	14,700	29,400
		12	17,600	14,080	35,200	30,500	24,900	17,600	35,200
	2	1	3100	2480	6200	5400	4400	3100	6200
		1½	4700	3760	9400	8100	6600	4700	9400
		1¾	5400	4320	10,800	9400	7600	5400	10,800
		2	6200	4960	12,400	10,700	8800	6200	12,400
		3	8800	7040	17,600	15,200	12,400	8800	17,600
		4	11,000	8800	22,000	19,100	15,600	11,000	22,000
		5	13,700	10,960	27,400	23,700	19,400	13,700	27,400
		6	16,500	13,200	33,000	28,600	23,000	16,500	33,000
		8	22,700	18,160	45,400	39,300	32,100	22,700	45,400
		10	28,400	22,720	56,800	49,200	40,200	28,400	56,800
		12	34,100	27,280	68,200	59,100	48,200	34,100	68,200
	4	1	5500	4400	11,000	9500	7800	5500	
		2	11,000	8800	22,000	19,100	15,600	11,000	
		3	16,400	13,120	32,800	28,400	23,200	16,400	
		4	20,400	16,320	40,800	35,300	28,800	20,400	—
		5	25,500	20,400	51,000	44,200	36,100	25,500	
		6	30,600	24,480	61,200	53,000	43,300	30,600	

* in lb
† in in.

Figure 6-7. The load rating of a web sling depends on the webbing class, number of plies, webbing width, and hitch type.

Fabrication efficiency is the ratio of the tensile strength of a webbing material to the tensile strength of the web sling into which it is fabricated. Fabrication efficiency accounts for the loss of strength in webbing due to stitching and other modifications. Typical fabrication efficiencies are 80% to 85% for single-ply slings but are lower for multiple-ply or very wide slings.

The safety factor indicates the fraction of total tensile strength that may be used for lifting loads. The typical safety factor is 5. For unusual circumstances, safety factors are increased to 8 or more to increase the margin of safety.

Web sling strength is based on new webbing material. Web sling strength decreases with age and mishandling, such as dragging the sling across a floor, tying it in knots, pulling it out from under a load when the load is resting on it, or dropping the metal fittings. Bunching webbing material between the ears of a clevis, shackle, or hook can also weaken web sling strength.

Web Sling Identification. All web slings must include a durable identification tag. **See Figure 6-8.** The tag must include the manufacturer name, model number, rated load for at least one hitch type (including the number of legs, if applicable, and sling angle), and type of webbing material. With this information, rated loads for other hitch types can be either calculated or found in a specification table. This tag must be sewn onto the webbing.

WEB SLING IDENTIFICATION TAGS

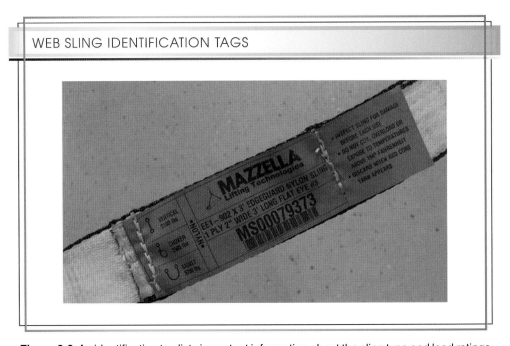

Figure 6-8. An identification tag lists important information about the sling type and load ratings.

Roundslings

A *roundsling* is an endless (continuous loop) sling made from a bundle of unwoven synthetic fiber yarns enclosed in a protective cover of synthetic fiber. **See Figure 6-9.** Roundslings make excellent choker-hitch slings because they are flexible and conform to the shape of the load. Also, due to their construction, choker hitches do not bind or lock up, which makes sling release simple.

ROUNDSLINGS

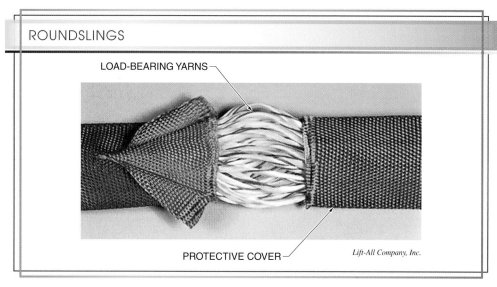

Lift-All Company, Inc.

Figure 6-9. A roundsling is composed of a bundle of unwoven yarns of load-bearing synthetic fibers enclosed in a non-load-bearing woven cover.

Roundsling Fabrication. The core yarns, typically polyester, are not woven but wound together with multiple turns. This winding is done uniformly to ensure even load-bearing distribution. As with webbing, brightly colored yarns are often included among the core yarns to serve as damage indicators.

The cover is made from polyester or nylon yarns woven into a continuous tubular shape. The cover is not load bearing, but it does protect the core yarns from damage, such as from abrasion or the environment. For example, the cover keeps damaging UV light from degrading the strength of the load-bearing core yarns.

Most roundslings are fabricated into a single large loop similar to Type V web slings. **See Figure 6-10.** These roundslings can be used in a variety of hitches. Additional sleeves may be placed over the roundslings to offer extra protection from abrasion or to create a sling configuration that has a loop eye at each end (eye-eye design). Manufacturers may also offer specialized slings based on roundsling materials and construction. For example, roundslings can be braided into a tough and extremely strong eye-eye type sling.

ROUNDSLING CONFIGURATIONS

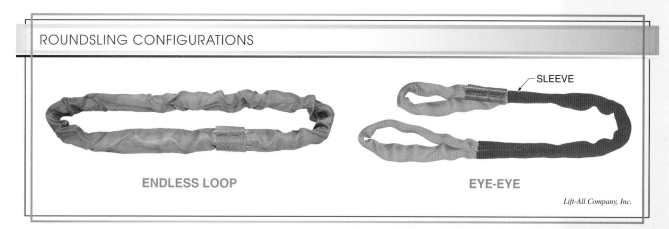

ENDLESS LOOP   EYE-EYE

Lift-All Company, Inc.

Figure 6-10. The primary roundsling configurations are the endless loop and the eye-eye types.

Roundslings are available in a variety of lengths. The length is the distance between points just inside the extreme ends when the sling is straightened to its full length. This includes any end fittings. Roundslings can stretch in length from about 2% to 5% when under load, so measurements are always taken when the sling is straightened but relaxed.

Roundslings can be used with typical metal rigging hardware. Some roundslings are manufactured with fittings permanently attached, such as hooks. For example, roundsling bridle hitches include two or more roundsling legs attached to a master link.

Roundsling Strength. Listed on roundsling specification tables are the rated loads for size and hitch configuration. **See Figure 6-11.** Sizes 1 to 13 are based on the *Recommended Standard for Synthetic Polyester Roundslings* developed by the Web Sling and Tie Down Association (WSTDA). The size number is related to a sling's relaxed diameter, so larger numbers correspond to stronger slings, but the number does not indicate an actual sling dimension. Alternatively, some manufacturers use their own size designations, though their sling specifications often match the standard.

RATED LOADS FOR ROUNDSLINGS

Size	Color Code	Vertical*	Choker*	Basket* Vertical	Basket* 60°	Basket* 45°	Basket* 30°
1	Purple	2600	2100	5200	4500	3700	2600
2	Green	5300	4200	10,600	9200	7500	5300
3	Yellow	8400	6700	16,800	14,500	11,900	8400
4	Tan	10,600	8500	21,200	18,400	15,000	10,600
5	Red	13,200	10,600	26,400	22,900	18,700	13,200
6	White	16,800	13,400	33,600	29,100	23,800	16,800
7	Blue	21,200	17,000	42,400	36,700	30,000	21,200
8	Orange	25,000	20,000	50,000	43,300	35,400	25,000
9	Orange	31,000	24,800	62,000	53,700	43,800	31,000
10	Orange	40,000	32,000	80,000	69,300	56,600	40,000
11	Orange	53,000	42,400	106,000	91,800	74,900	53,000
12	Orange	66,000	52,800	132,000	114,300	93,000	66,000
13	Orange	90,000	72,000	180,000	155,900	127,300	90,000

* in lb

Figure 6-11. The rated loads of a roundsling depend on its size and hitch type. Roundslings are often color coded for easy identification.

Roundslings are particularly useful as choker hitches, but the configuration of this hitch has a significant effect on the strength of the sling. For example, if the choke is drawn tightly to the body of the load, part of the sling will be highly bent, which reduces its strength. Conversely, if the choke is high above the load, the choke angle will be greater, and the sling will retain more of its strength. A table of choke angles can be used to determine the derating of a rated load for a roundsling used as a choker hitch. **See Figure 6-12.**

ROUNDSLING CHOKER HITCH DERATING

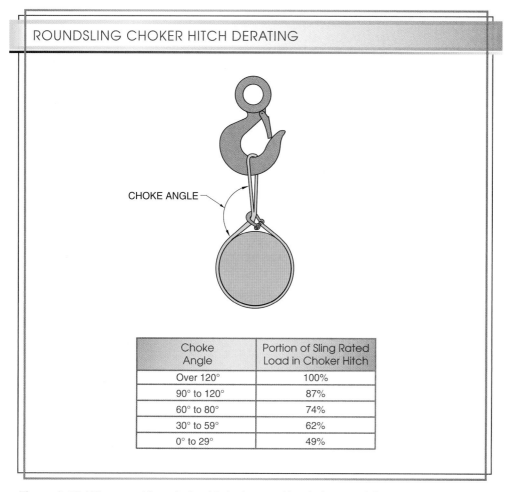

CHOKE ANGLE —

Choke Angle	Portion of Sling Rated Load in Choker Hitch
Over 120°	100%
90° to 120°	87%
60° to 80°	74%
30° to 59°	62%
0° to 29°	49%

Figure 6-12. When used in a choker hitch, the rated load of a roundsling must be derated for choke angles up to 120°.

Roundsling Identification. The rated load of a roundsling is listed on its identification tag and indicated by the color of its cover. The color codes may vary by manufacturer, though most follow a common sequence specified by the WSTDA. However, the color code of a sling should not be relied on to determine its load rating. The identification tag should always be referred to for the load rating of a sling.

Identification tags must include certain information, such as the name of the manu-facturer, the model number, rated capacity for at least one hitch type and its angle (if applicable), core material, cover material (if different from core material), and number of legs (if a bridal sling assembly). **See Figure 6-13.** Beyond this required information, it is recommended that the tag include the rated loads for at least three basic hitches (for single-leg slings) or uses at 60°, 45°, and 30° angles (for multiple-leg slings). While the rated load for a single hitch can be used with calculations or tables to determine the rated loads of most other hitches, it is more convenient to have multiple rated load specifications readily available. It is also recommended that the tag indicate the sling's length, which can be used to determine if the sling has stretched from overloading.

ROUNDSLING IDENTIFICATION TAGS

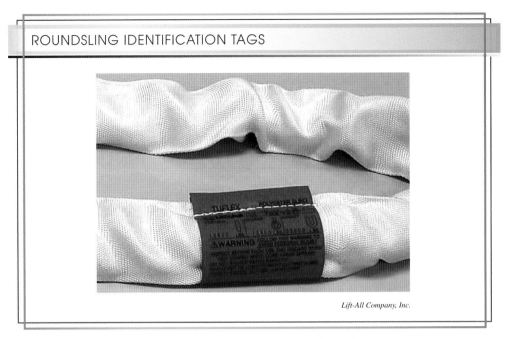

Lift-All Company, Inc.

Figure 6-13. A roundsling tag lists important identifying information.

Synthetic Sling Inspection

Use over time can obscure the print on an identification tag or remove the tag entirely from the sling. So the first part of an inspection is to find the tag. **See Figure 6-14.** If a tag is damaged or missing but the sling appears acceptable, the sling may be returned to the manufacturer for testing and retagging.

DAMAGED IDENTIFICATION TAGS

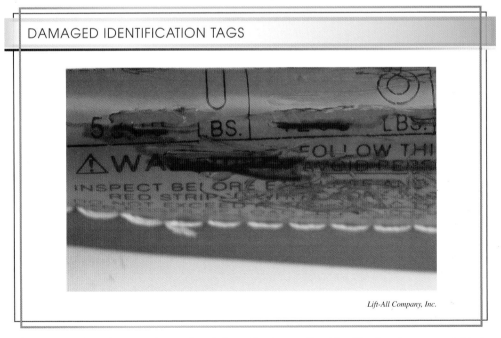

Lift-All Company, Inc.

Figure 6-14. Any damage that makes information on a sling identification tag unreadable is reason to immediately remove the sling from service.

Synthetic slings generally include red warning yarns within the body that are not normally visible. These yarns are designed to be visible if the sling material is damaged sufficiently to render it unfit for use. **See Figure 6-15.** A sling must be immediately removed from service if the warning yarns are visible. However, a sling is not necessarily safe to use if the warning yarns are not visible. Some types of damage may not expose these yarns but can still weaken the sling, such as damage from UV exposure. Furthermore, roundslings may not contain warning yarns, but the slings must be discarded if any of the interior load-bearing yarns are visible.

EXPOSED WARNING YARNS

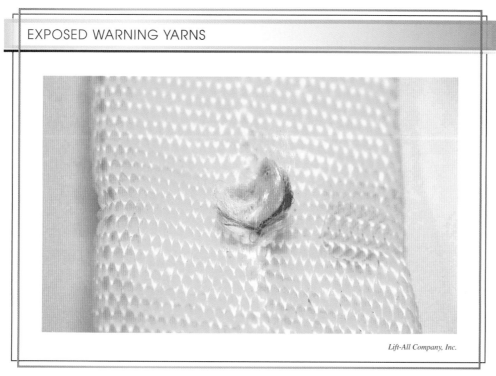

Lift-All Company, Inc.

Figure 6-15. Red warning yarns are embedded inside synthetic slings. They are normally hidden but become exposed due to some types of damage.

Slings should be inspected for evidence of physical, thermal, or chemical damage. **See Figure 6-16.** Excessive wear or abrasion can damage the stitching in load-bearing splices or cause the surface fibers to fray and break. Significant cuts or tears are those that damage 50% of the longitudinal yarns (those running along the length of the sling) in an area that spans ¼ of the webbing width or 100% of the yarns for ⅛ of the width. Edge cuts can be particularly weakening. Other types of physical damage include punctures, snags, and embedded particles.

Thermal damage, which can be caused by heat or friction, is indicated by melted or charred webbing material. Exposure to certain chemicals can cause disintegration of the synthetic fibers, depending on the fiber material. This disintegration may appear as discoloration, brittleness, and/or stiffness in areas of the sling.

Any attached hooks or other rigging hardware should also be inspected for damage in accordance with established hardware inspection procedures. A sling assembly should be removed from service if any attached hardware shows excessive corrosion, pitting, gouging, twisting, or bending.

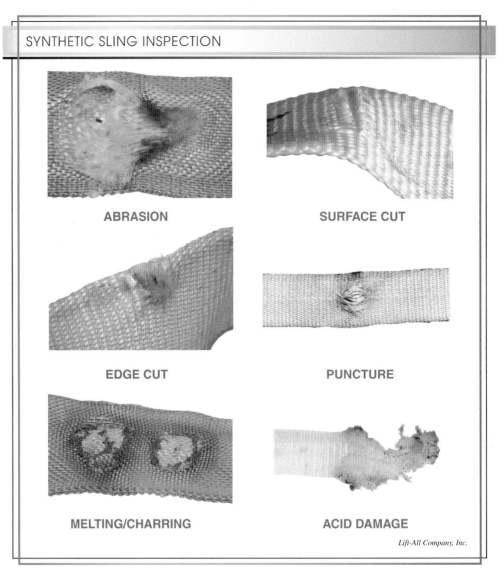

SYNTHETIC SLING INSPECTION

ABRASION

SURFACE CUT

EDGE CUT

PUNCTURE

MELTING/CHARRING

ACID DAMAGE

Lift-All Company, Inc.

Figure 6-16. Web slings and roundslings are weakened by many types of physical, thermal, and chemical damage.

Web sling material is relatively strong and may be used to lift large loads.

Another serious problem with both web slings and roundslings is knotting. **See Figure 6-17.** A knot alters the load-bearing characteristics of the yarns at that location, significantly weakening the sling. Knots can be easily introduced when sorting long slings. Knots should be untied only if they have not yet been pulled tight under load. If not untied before a sling is loaded, the knots in the sling are often irreversibly tightened, which can also cause damage to the fibers. A sling with any tightened knot must be removed from service.

KNOTS

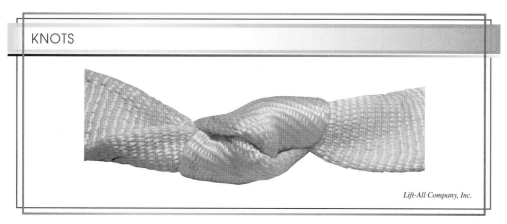

Lift-All Company, Inc.

Figure 6-17. If a knot has been pulled tight, the strength of the sling may be permanently compromised.

CHAIN SLINGS

Chain slings are fabricated in single-leg and multiple-leg configurations. Multiple-leg configurations consist of two, three, or four legs. **See Figure 6-18.** Multiple-leg slings are also known as bridle slings.

RATED LOADS* OF CHAIN SLINGS

Class	Nominal Chain Size†	Single-Leg Vertical	Two-Leg Sling			Four-Leg Sling		
			60°	45°	30°	60°	40°	30°
80	7/32	2100	3600	3000	2100	5500	4400	3200
	9/32	3500	6100	4900	3500	9100	7400	5200
	5/16	4500	7800	6400	4500	11,700	9500	6800
	3/8	7100	12,300	10,000	7100	18,400	15,100	10,600
	1/2	12,000	20,800	17,000	12,000	31,200	25,500	18,000
	5/8	18,100	31,300	25,600	18,100	47,000	38,400	27,100
	3/4	28,300	49,000	40,000	28,300	73,500	60,000	42,400
	7/8	34,200	59,200	48,400	34.200	88,900	72,500	51,300
	1	47,700	82,600	67,400	47,700	123,900	101,200	71,500
	1¼	72,300	125,000	102,200	72,300	187,800	153,400	108,400
100	7/32	2700	4700	3800	2700	7000	5700	4000
	9/32	4300	7400	6100	4300	11,200	9100	6400
	5/16	5700	9900	8100	5700	14,800	12,600	8500
	3/8	8800	15,220	12,400	8800	22,900	18,700	13,200
	1/2	15,000	26,000	21,200	15,000	39,000	31,800	22,500
	5/8	22,600	39,100	32,000	22,600	58,700	47,900	33,900
	3/4	35,300	61,100	49,900	35,300	91,700	74,900	53,000
	7/8	42,700	74,000	60,400	42,700	110,900	90,600	64,000

* in lb
† in in.

Figure 6-18. Rated loads of chain slings depend on the chain grade, size, number of legs, and sling angle.

Chain slings are manufactured from lengths of chain and rigging hardware components. Single-leg slings are constructed of a single length of chain and may have a hook or a master link at one or both ends. Multiple-leg slings have multiple chain legs attached to a single master link and hooks attached to the end of each leg. A variety of hook types are available. Sling assemblies cannot typically be modified by the user or in the field because the hardware is permanently attached. Therefore, they must be ordered premade in the required configuration. **See Appendix.**

After a manufacturer assembles a chain sling, it must be subjected to a proof test. A *proof test* is a nondestructive test in which a sling is subjected to a tension force greater than its rated load but less than its breaking strength. A single-leg sling is tested at twice its rated load. Master links are tested at four times the rated load of an individual leg for double-leg bridle slings or six times the rated load for three- and four-leg bridle slings.

Documentation of the proof test is provided with each new sling, and OSHA requires that this proof test verification be retained by the employer. If a sling must be repaired, it must again be tested to confirm its integrity, and the new proof test record must be retained.

Chain Sling Identification

Chain sling specifications are identified by the manufacturer on an attached nameplate or tag. **See Figure 6-19.** This tag includes the name of the manufacturer, the grade of chain, chain size, number of legs, rated load for at least one hitch type, serial number, and reach. The *reach* is the distance from the inside of a sling's upper end fitting to the inside of its lower end fitting. This length is significant when making subsequent inspections of the sling to determine whether the sling has elongated.

CHAIN SLING IDENTIFICATION TAGS

Figure 6-19. Chain slings have durable identification tags that list important sling type and rated load information.

Grade 80 and Grade 100 chains must also include identification directly on the chain links. **See Figure 6-20.** Identifying numbers are embossed on links at intervals no greater than 3′ apart. The characters are raised and include a manufacturer's mark, traceability or date code, and grade indicator. Grade 80 chain is indicated by "8," "80," or "800." Grade 100 chain is indicated by "10," "100," or "1000."

GRADE MARKINGS

Figure 6-20. Chain grades must be stamped on links at certain intervals.

Chain Sling Applications

Chain has little friction against most materials, so it has a tendency to slip when wrapped around loads, such as in a basket or choker hitch. Temporary stops may have to be applied to a load, or a lifting beam may have to be utilized to prevent slippage of the slings.

Chain slings should never be shortened by twisting or by using bolts or other makeshift fasteners. Special fittings are available that allow the lengths of chain legs to be adjusted, but care should be taken to ensure that the links of the chains are properly inserted or hooked into the fittings. **See Figure 6-21.** As with other multiple-leg slings, sling load calculations should be performed to ensure that the individual legs of a sling are not overloaded, especially when connected to asymmetrical loads.

Protective gloves should be worn when handling chain slings, and caution should be taken to ensure that body parts are not pinched by the links or components of a sling as a load is lifted. Chain slings should be checked for twisting or kinking as they are placed under tension. It may be necessary to disconnect a sling to remove any twists before hoisting a load.

SHORTENERS

The Crosby Group LLC

Figure 6-21. Chain sling legs may be shortened only with hardware approved for this use by the chain manufacturer or a qualified person.

WIRE ROPE SLINGS

Wire rope is used extensively with hoists and cranes of all types, and it is also used to fabricate slings. Wire rope slings are strong, durable, and relatively flexible, making them a commonly used sling type for a variety of rigging applications. **See Figure 6-22.**

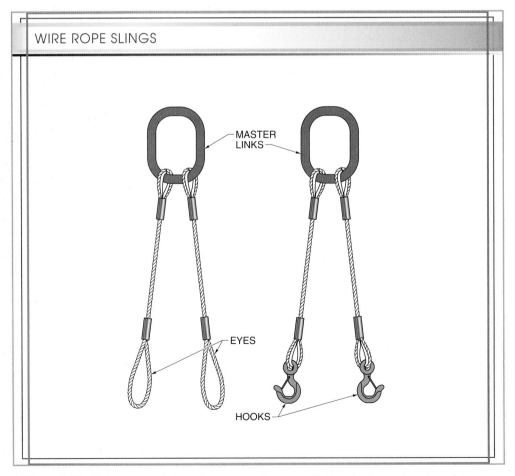

WIRE ROPE SLINGS

MASTER LINKS

EYES

HOOKS

Figure 6-22. Slings are made from lengths of wire rope with terminations at each end.

Wire rope slings are fabricated from a length of wire rope. The rope must first be cut, and then the two ends are terminated in some way to allow the rope to be used with other rigging hardware. Terminations involve either forming and securing an eye with a loop of rope or attaching an end fitting of the desired type. Slings can incorporate a combination of any two types of terminations, depending on the rigging requirements.

Terminations affect the rated load of a sling because the necessary bends may weaken the rope and the fastenings may not have the same strength as the rope. *Termination efficiency* is the ratio of the rated load of a wire rope sling to the rated load of the unterminated wire rope. For example, a termination with 90% efficiency reduces the rated load of the sling to 90% of the rated load of the wire rope.

Wire rope slings are commonly purchased prefabricated with the desired rope size, length, and end fittings. However, with the proper tools, qualified personnel can also fabricate custom wire rope slings.

Seizing and Cutting

Before cutting, a rope end must be bound to prevent strand unraveling or unsafe loose wires. *Seizing* is the wire wrapping that binds the end of a wire rope near where it is cut. **See Figure 6-23.** The binding holds the strands firmly in place. Adequate binding prevents rope distortion, flattening, or loosening of strands while making terminations. Inadequate binding may allow uneven distribution of the load on the strands during lifting, which reduces strength and shortens rope life.

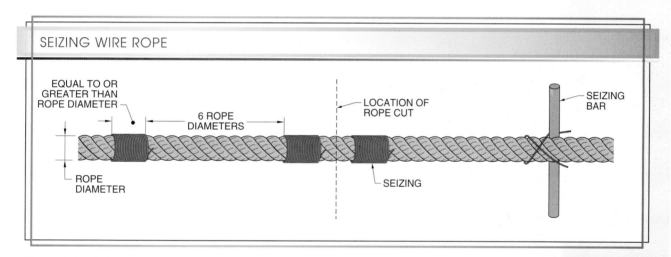

Figure 6-23. Seizing is wire wrapping added to wire rope to prevent unraveling or loose wires after cutting.

Normally, one seizing on each side of the planned cut is sufficient for preformed wire rope. *Preformed rope* is wire rope in which the strands are permanently formed into a helical shape during fabrication. The wires in preformed ropes do not easily unravel when cut. Common wire ropes, those that are not preformed or are rotation resistant, normally require a minimum of two seizings on each side of the cut, which are placed six rope diameters apart. Other seizing requirements vary based on rope size. Always check the manufacturer's specifications for recommendations.

The typical method for seizing a wire rope is to lay one end of the seizing wire between two strands of the wire rope. A seizing bar is used to wind the other end of the seizing wire tightly around the rope, without overlapping, until the required seizing width is obtained. The seizing is secured by twisting the ends of the seizing wire together.

The wire rope is then cut using a rope shear, an abrasive cutoff wheel, or an oxyacetylene cutting torch. Shearing or abrasive cutting leaves a sharp edge that should be filed smooth. An oxyacetylene cutting torch is preferred because the heat also fuses the strands and strand wires together.

Eye Terminations

An eye is fabricated by turning the end of the cable back on itself so that it forms a loop with a dead end. **See Figure 6-24.** The *dead end* is the loose end of a rope. The *live end* is the load-lifting end of a rope. A thimble is often used to support the loop. The loop must then be secured by binding the dead and live ends together at the base of the eye. Most slings use swaged sleeves, particularly prefabricated slings. U-bolt clips can also be used to form an eye in wire rope, but OSHA generally prohibits this method for fabricating slings.

EYE TERMINATIONS

Lift-All Company, Inc.

Figure 6-24. Eye terminations are formed by turning the end of the cable back on itself so it forms a loop.

Thimbles. A *thimble* is a curved piece of metal that supports a loop of rope and protects it from sharp bends and abrasion. **See Figure 6-25.** A thimble prevents damage to a wire rope by both preventing wear on the inside of the eye and maintaining the diameter of the loop. Thimbles may also be used with natural fiber or synthetic fiber rope.

Thimbles are made from formed or cast metal. They must be sized according to the diameter of the rope being used. Proper sizing is important because a thimble that is too small causes damage by pinching the rope, and a thimble that is too large does not give the proper side support to the rope.

THIMBLES

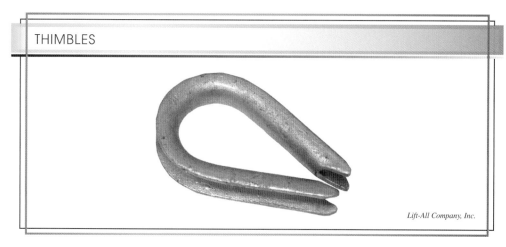

Lift-All Company, Inc.

Figure 6-25. A thimble is a metal guard that protects the inside surface of a rope eye from abrasion and tight bending.

Swaged Sleeves. A *swaged sleeve* is a compression fitting that is crimped onto the two portions of wire rope that meet at the base of an eye loop. **See Figure 6-26.** The sleeve is threaded onto the end of a rope before the eye loop is formed. Then, the rope end is inserted back into the sleeve, forming the loop. After inserting a thimble, if any, and adjusting the size of the loop, the sleeve is swaged, securing the eye loop. *Swaging* is a mechanical process of forming a metal into a certain shape, usually at ambient temperatures.

SWAGED SLEEVES

SWAGED SLEEVES

Figure 6-26. Swaged sleeves are fittings that are crimped onto the bases of wire rope eyes.

The sleeve is made from a material that is strong, yet soft enough to deform during installation and make a tight mechanical bond to the rope. A special crimping tool is used to compress the sleeve, causing the inside surface to mold around the shape of the rope. This is an effective method of fastening the rope parts together. The efficiency rating is between 90% and 100%.

U-Bolt Clips. U-bolt clips, also called wire rope clips, are used to hold two parts of a rope together with a nut-and-bolt-like fastener. A U-bolt clip consists of a saddle, a threaded U-bolt, and two nuts. The clip should be assembled with the U-bolt in contact with the dead end of the wire rope and the saddle in contact with the live end. Otherwise, the rope could be damaged and weakened if the U-bolt is tightened onto the live end of the rope.

Only wire rope clips that are manufactured specifically for rigging and hoisting applications should be used. These clips have forged steel saddles and are many times stronger than non-load-rated clips, which have cast iron or zinc saddles.

Clips must be arranged, spaced, and assembled properly to maintain the strength of a rope. **See Figure 6-27.** The required turnback length and number of clips are determined by the rope size and/or the manufacturer's specifications. The *turnback* is the portion of the end of a rope that is folded back on itself.

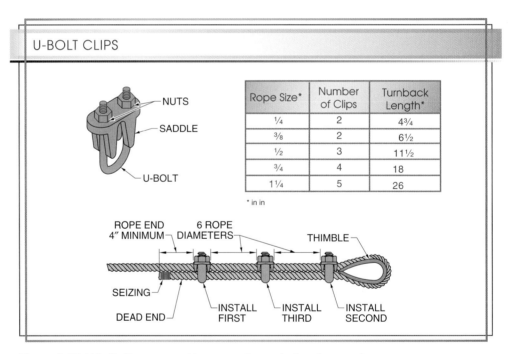

U-BOLT CLIPS

Rope Size*	Number of Clips	Turnback Length*
¼	2	4¾
⅜	2	6½
½	3	11½
¾	4	18
1¼	5	26

* in in

Figure 6-27. U-bolt clips are used to secure the end of a wire rope in an eye.

The first clip is placed near the end of the turnback, and the nuts are tightened alternately with a torque wrench to the manufacturer's recommended torque value. The second clip is placed at the end of the thimble, and the nuts are firmly tightened but not fully torqued. Any other clips are spaced evenly between the first two and torqued. Finally, a load is placed on the rope, and the nuts of every clip are retorqued. The efficiency of a termination made with clips is approximately 80%.

When in use, wire rope clips must not be in contact with the load or with other obstructions. When performing periodic inspections of wire rope eyes with clips, the nuts should be retightened to the manufacturer's specifications.

Socket Terminations

A *socket* is a fitting attached to the end of wire rope to provide a means for making strong connections. Sockets provide closed or open fittings to which other rigging hardware can be attached. A closed fitting is a solid loop. An open fitting has a part that can be removed in order to attach it to a closed fitting.

Wire rope sockets include swage, spelter, and wedge designs. Swage and spelter sockets are permanent wire rope attachments and have the highest efficiency ratings at 95% to 100%. Wedge sockets can be installed and removed as needed. Their efficiencies are around 80%.

Swage Sockets. A *swage socket* is a socket that is compressed onto the end of a wire rope. **See Figure 6-28.** As with swaged sleeves, a swage socket must be compressed in a hydraulic press to achieve the necessary binding to a rope. The inside of the socket conforms to the shape of the rope strands and locks the socket into place.

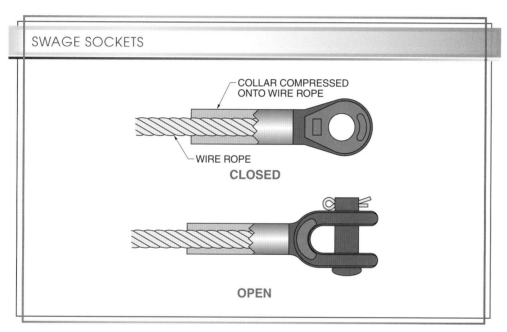

SWAGE SOCKETS

COLLAR COMPRESSED ONTO WIRE ROPE

WIRE ROPE

CLOSED

OPEN

Figure 6-28. A swage socket is compressed with high pressure onto the end of a wire rope until the collar conforms to the shape of the rope.

Spelter Sockets. A *spelter socket* is a socket that uses molten zinc or resin to secure the end of a wire rope inside the socket. **See Figure 6-29.** A rope end is inserted through a socket collar, and the individual wires are separated and fanned out. Molten zinc or resin is poured into the collar and hardens around the wires, creating a solid wedge-shaped assembly that resists sliding back through the socket.

Wedge Sockets. A *wedge socket* is a socket that holds a loop of wire rope securely with a wedge that is tightened by tension on the rope. **See Figure 6-30.** Wedge sockets are popular because they can be installed and repositioned quickly and easily. However, due to its design, a wedge socket can be installed incorrectly, creating a sharp bend on the live end of a rope. The live end must be aligned with the socket. The exposed dead-end section must extend out of the wedge a minimum of eight rope diameters.

SPELTER SOCKETS

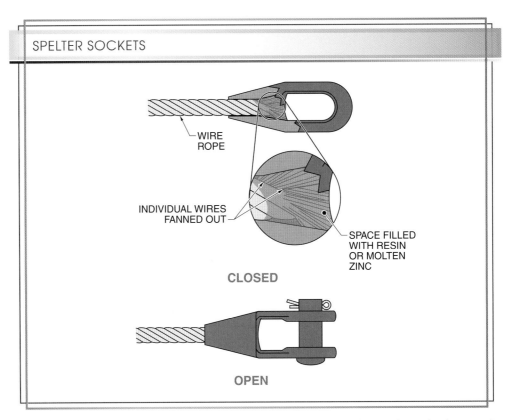

Figure 6-29. A spelter socket is secured to the end of a wire rope when resin or zinc is poured over a fanned-out wedge of wire, locking the rope in place.

WEDGE SOCKETS

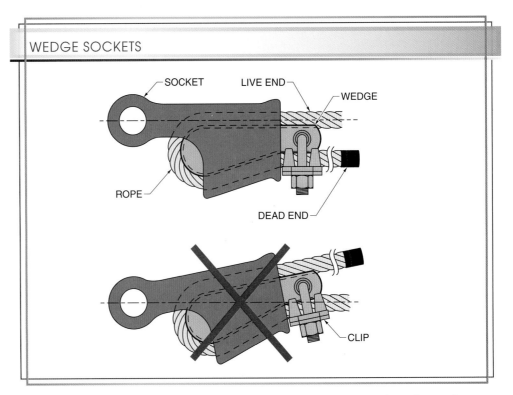

Figure 6-30. A wedge socket holds tightly to a wire rope when the rope is under tension.

COMMON APPLICATIONS—
REPLACING POLE-MOUNTED TRANSFORMERS WITH CAPSTAN HOISTS

A capstan hoist is an electric winch that is used to lift heavy loads by wrapping rope around the rotating drum of the hoist. A capstan hoist may be used to replace a pole-mounted transformer in areas where a digger derrick or other large lifting equipment cannot access the work area. The capstan hoist is typically mounted to the base of a utility pole.

To replace a pole-mounted transformer with a capstan hoist, apply the following procedure:

1. Install a transformer gin at the top of the utility pole.
2. Attach a capstan hoist near the base of the utility pole.
3. Attach a transformer sling to the lifting points on the transformer.
4. Attach the transformer sling to the winch line.
5. Activate and use the capstan hoist to lift the transformer into position.
6. Secure the transformer to the utility pole with the proper bolts or brackets.

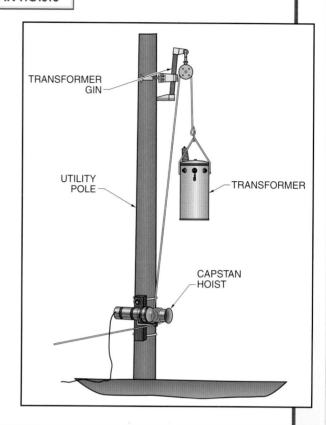

TRANSFORMER GIN

UTILITY POLE

TRANSFORMER

CAPSTAN HOIST

Chapter 6 Learner Resources

Lineworker Rigging Practices

LOADS AND FORCES 7

OBJECTIVES

- Calculate load weight based on material and surface area or volume.
- Describe the importance of a load's center of gravity on rigging.
- Estimate the position of the center of gravity of a load using various procedures.
- Describe the effect of rigging symmetry on force distribution.
- Describe the different types of force on a load.
- List and describe the different types of mechanical advantage.
- Explain what block and tackle are used for.
- Explain why reeving and block loading are important.

Rigging a load for safe hoisting requires knowledge of its weight and center of gravity. This knowledge is necessary to ensure that the load will remain balanced and manageable and that the rigging components are strong enough to support the load weight. Load weight and center-of-gravity information are sometimes provided, but often they must be determined by the rigger. Simple procedures and calculations can be used to make reasonable estimations.

LOAD WEIGHT

The most critical factor in rigging and lifting is load weight. The weight must be determined when planning a lift so that the proper rigging and hoisting equipment can be selected. The lifting capacity of each rigging component must be adequately rated for at least its portion of the load weight. **See Figure 7-1.**

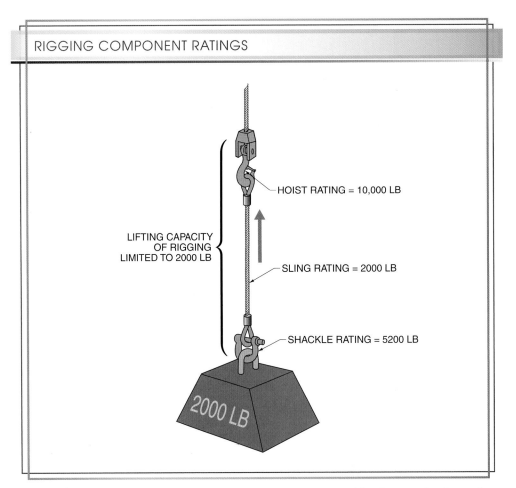

RIGGING COMPONENT RATINGS

HOIST RATING = 10,000 LB

LIFTING CAPACITY OF RIGGING LIMITED TO 2000 LB

SLING RATING = 2000 LB

SHACKLE RATING = 5200 LB

2000 LB

Figure 7-1. Each component in a rigging assembly must be capable of supporting its portion of the load weight.

The load weight must also be known because certain areas of the destination's structure may not be able to support the load. Placing the load in these areas could be extremely hazardous and must be avoided. Any weight limitations should be considered when developing a lifting plan.

Load weight is sometimes indicated on equipment nameplates, on shipping documents, or in the manufacturer's product information. **See Figure 7-2.** However, the load weight may be greater than shown in the original data if the load has been modified, is inside a shipping container, or is on a skid. Verifying that the weight has not changed since the documentation was prepared or that the weights of any additions to the load have been accounted for is also part of developing a lifting plan.

If a load's weight is unknown, it can be calculated by using either stock material weight tables or by using area, volume, and material weight information. The weight of the rigging equipment may also need to be added to the load weight if it could be a significant portion of the overall load weight. Relatively small hardware can be weighed with a scale, and larger equipment, such as a spreader bar, is usually marked with its weight. The total, overall weight is used to ensure that all equipment, including the crane or hoist, is properly rated. This additional weight likely has a negligible effect of the load's center-of-gravity (CG) location, though, and can usually be ignored for CG calculations.

EQUIPMENT NAMEPLATES

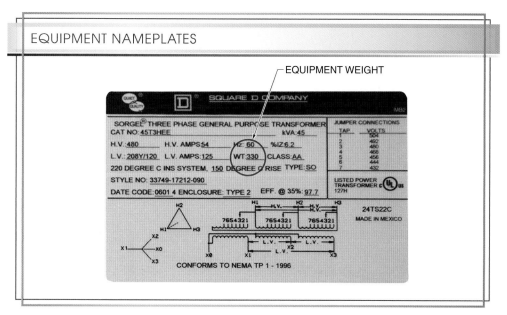

Figure 7-2. The weight of a load, particularly of packaged equipment, can sometimes be found on nameplates.

Stock Material Weight Tables

Stock material weight tables can be used when a load consists of common stock or structural shapes, such as round or square bars, round or square tubing, I-beams, angles, tees, channels, or plates. These shapes may be available in steel, aluminum, copper, brass, or other materials. Stock material weight tables list the weight of materials by their linear, area, or volumetric measurements in either English or metric units. **See Appendix.** Tables for common shapes and sizes are readily available in material reference books or in manufacturer literature. **See Figure 7-3.** For example, a 1″ diameter, round steel bar weighs 2.67 lb/ft.

The weight of linear stock material is calculated by applying the following formula:

$$W_{total} = n \times l \times W_l$$

where

W_{total} = total weight (in lb)

n = number of pieces

l = length (in ft)

W_l = linear unit weight (in lb/ft)

For example, what is the total weight of 10 utility poles that are 40′ long? *Note:* The utility poles weigh 30 lb/ft.

$$W_{total} = n \times l \times W_l$$
$$W_{total} = 10 \times 40 \times 30$$
$$W_{total} = \textbf{12,000 lb}$$

WEIGHT OF SELECTED STEEL STOCK MATERIALS			
Diameter or Thickness*	Round Bar†	Square Bar†	Sheet/Plate†
1/16	—	—	2.55
1/8	0.0417	0.0531	5.11
3/16	0.0939	0.120	7.66
1/4	0.167	0.213	10.2
3/8	0.376	0.478	15.3
1/2	0.668	0.850	20.4
3/4	1.50	1.91	30.6
1	2.67	3.40	40.8
1 1/4	4.17	5.31	51.1
1 1/2	6.01	7.65	61.3
1 3/4	8.18	10.4	71.5
2	10.7	13.6	81.7
4	42.7	54.4	163
6	96.1	122	245

* in in.
† in lb/ft
† in lb/ft²

Figure 7-3. The weight of a load consisting of stock materials can be estimated using the unit weights in stock material weight tables.

Load weights of stock plates or sheets are calculated by multiplying the material unit weight (from stock material weight tables) by the number of plates or sheets and the area of a single plate or sheet, as shown in the following formula:

$$W_{total} = n \times l \times w \times W_A$$

where

W_{total} = total weight (in lb)

n = number of pieces

l = length (in ft)

w = width (in ft)

W_A = area unit weight (in lb/ft²)

For example, what is the total material weight of a load consisting of 35 pieces of 4′ × 8′ steel sheet that is 1/16″ thick? *Note:* A steel sheet at 1/16″ thick weighs 2.55 lb/ft².

$$W_{total} = n \times l \times w \times W_A$$
$$W_{total} = 35 \times 8 \times 4 \times 2.55$$
$$W_{total} = \textbf{2856 lb}$$

Material Weight Calculations

If a load does not consist of raw stock shapes and its weight information is not available, the weight can be estimated using the load size (area and/or volume) and the material's unit weight. This weight information can be found in many types of references. For example, density is weight per unit volume. **See Figure 7-4.**

If the load is made from a relatively thin plate or sheet, its surface area is multiplied by the area unit weight of the plate or sheet in order to determine the load weight. This is the same concept as estimating weight from the surface area of stock shapes. However, the area calculations are typically more complicated because the load is not simply a stack of identical stock pieces. Instead, the load may be a fabricated part that consists of multiple pieces assembled together.

COMMON MATERIAL DENSITIES

Material	lb/ft³	lb/in³
Steel	490	0.284
Aluminum	165	0.0955
Concrete	150	0.0868
Wood	50	0.0289
Water	62	0.0359
Sand and gravel	120	0.0694
Copper	560	0.324
Oil	58	0.0356

Figure 7-4. The density of a material, along with load measurements, can sometimes be used to estimate load weight.

Determining the total surface area of the load may involve calculating the area of each individual piece of the load separately. Usually, each piece is a simple shape with an easily calculated area. **See Figure 7-5.** For example, one piece may be rectangular, another may be circular, and so on. The surface area of each simpler shape is then added together. The sum is the surface area of the entire load. The load weight is then calculated using the following formula:

$$W_{total} = A \times W_A$$

where

W_{total} = total weight (in lb)

A = surface area (in ft²)

W_A = area unit weight (in lb/ft²)

If a load is made from thick, solid parts, its volume is multiplied by the material's density (weight per unit volume) in order to determine the load weight. Some loads are simple shapes, such as a block of steel, and some are complex shapes. Like the surface area calculation, complex shapes can usually be considered as a collection of simpler three-dimensional shapes, such as cubes or cylinders, whose volumes can be calculated individually. The sum is the volume of the entire load. The load weight is then calculated using the following formula:

$$W_{total} = V \times W_V$$

where

W_{total} = total weight (in lb)

V = volume (in ft³)

W_V = density (in lb/ft³)

For example, a load consists of a wind turbine tower section. The tower section is 40′ long, is 12′ in diameter, and has a wall thickness of 1″. What is the total weight of the load?

AREAS AND VOLUMES OF SIMPLE SHAPES

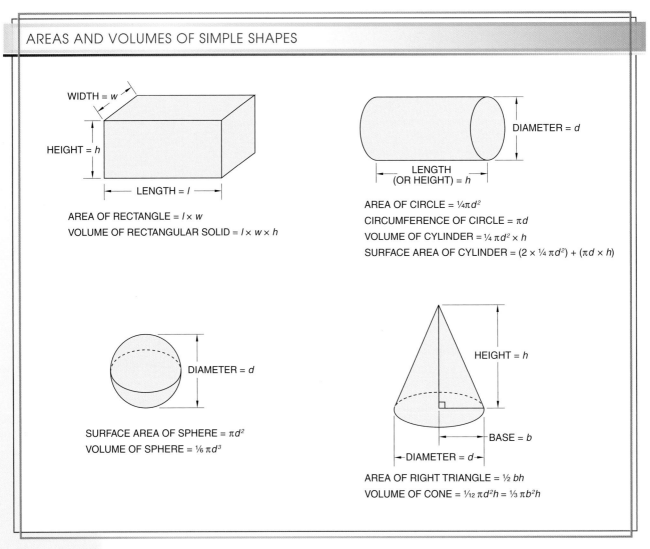

Figure 7-5. The shapes of some loads can be broken down into simpler shapes whose volumes or surface areas can be used to estimate load weight.

The weight of the tower section is calculated using the tower diameter. Since the section is hollow, the tower section is calculated as a solid object separately from the inside (hollow) section. **See Figure 7-6.** Then the hollow section is subtracted from the solid tower section to determine the total load weight. Since the tower section is made from steel, its weight can be estimated by calculating the volume of the section and multiplying by the density of steel (490 lb/ft³). The volume of the tower is calculated using the following formula:

$$V_{cylinder\,A} = \frac{\pi d^2}{4} h$$

where

$V_{cylinder\,A}$ = volume of solid cylinder (in ft³)

$\pi = 3.14$

d = diameter (in ft)

h = height of cylinder (in ft)

therefore

$$V_{cylinder\ A} = \frac{\pi d^2}{4} h$$

$$V_{cylinder\ A} = \frac{3.14 \times 12^2}{4} \times 40$$

$$V_{cylinder\ A} = \frac{452.16}{4} \times 40$$

$$V_{cylinder\ A} = 113.04 \times 40$$

$$V_{cylinder\ A} = \textbf{4521.6 ft}^3$$

The volume of the hollow section is calculated using the same formula as follows:

$$V_{cylinder\ B} = \frac{\pi d^2}{4} h$$

$$V_{cylinder\ B} = \frac{3.14 \times 11.83^2}{4} \times 40$$

$$V_{cylinder\ B} = \frac{439.45}{4} \times 40$$

$$V_{cylinder\ B} = 109.86 \times 40$$

$$V_{cylinder\ B} = \textbf{4394.4 ft}^3$$

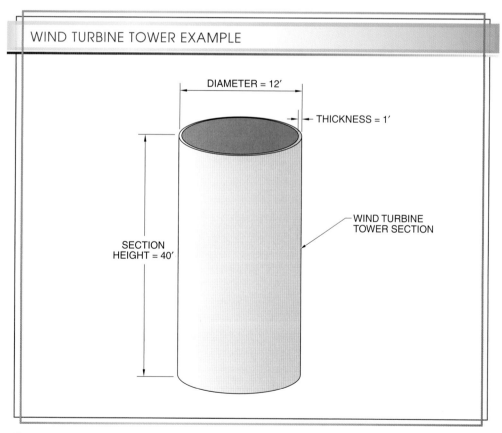

WIND TURBINE TOWER EXAMPLE

DIAMETER = 12′

THICKNESS = 1′

WIND TURBINE
TOWER SECTION

SECTION
HEIGHT = 40′

Figure 7-6. The weight of a cylindrical steel tank can be estimated using the formula for the surface area of a cylinder.

The volume of the hollow section (4394.4 ft³) is subtracted from the volume of the solid tower section (4521.6 ft³). The actual volume of the tower section is calculated using the following formula:

$$V = V_{cylinder A} - V_{cylinder B}$$

where

V = volume of tower section

$V_{cylinder A}$ = volume of solid cylinder (ft³)

$V_{cylinder B}$ = volume of hollow cylinder (ft³)

therefore

$V = 4521.6 - 4394.4$

$V = 127.2 \text{ ft}^3$

The weight of the tower is then determined by multiplying the volume by the density.

$$W_{total} = V \times W_V$$

$$W_{total} = 127.2 \times 490$$

$W_{total} = 62,328 \text{ lb}$

Therefore, the wind turbine tower section weighs about 62,328 lb.

LOAD BALANCE

Rigging must be arranged so a load remains stable and level when lifted. Any shifting, tipping, or rocking of a load may cause rigging to fail, allowing the load to fall and cause damage or injury. Keeping a load under control requires determining the load's center of gravity.

Center of Gravity

The *center of gravity (CG),* also known as the center of mass, is the point in space at which an object's mass is considered to be concentrated. **See Figure 7-7.** Load weight is considered to be located at this single point for all calculations and relationships. The CG is the balancing point of a load. Lifting a load from directly over its CG puts the least stress on the rigging and provides the safest and most controlled lifting conditions.

The CG is usually located within the body of a load, often near its physical center. The load's distribution of mass determines the exact CG. **See Figure 7-8.** That is, if one side of the load is larger or made from heavier materials, the CG will be closer to that end. It should also be noted that CG is a location in three-dimensional space. For example, the CG of a load may be in the center of the load from left to right but off-center from front to back.

Sometimes the appearance of a load indicates how evenly the mass is distributed. If the load is symmetrical along one dimension, then the CG is likely to be in the middle of that dimension. *Symmetry* is the characteristic of one side of an object mirroring its opposite side. However, symmetry is only a guideline and should not be relied on as the sole means of locating the CG of a load. Hidden, asymmetrical internal features can significantly affect the location of the CG.

The CG of a load must be determined before hoisting it, though this can be difficult with complex assemblies. Equipment manufacturers often mark the CG on their products or include specification sheets with rigging and/or CG information. **See Figure 7-9.** The CG can also be calculated precisely from weight and material information, but this can be time-consuming. In many cases, a load's CG can be determined by a simple trial-and-error procedure, but only if lifting an unbalanced load slightly will not create a safety hazard due to the load shifting. If these options are not possible or practical, some simple procedures can be used to estimate the CG position.

CENTER OF GRAVITY (CG)

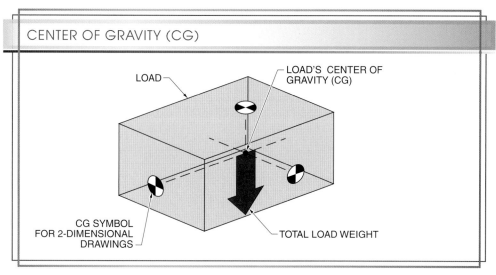

Figure 7-7. The CG of a load is the single point in space at which an object's mass is considered to be concentrated. A symbol is used to mark the location on two-dimensional drawings.

CG LOCATIONS

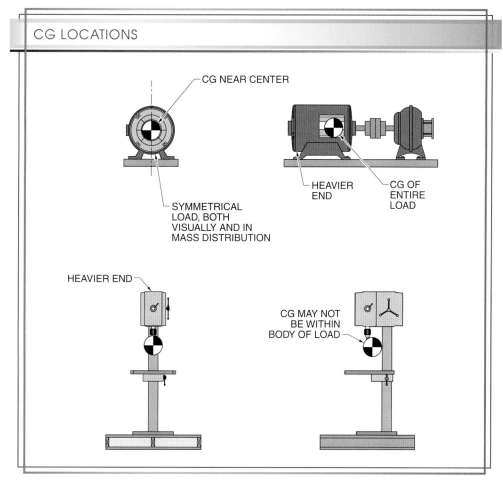

Figure 7-8. The location of a load's CG is often related to the load's symmetry. However, in asymmetrical loads, the CG is always closer to the heavier end.

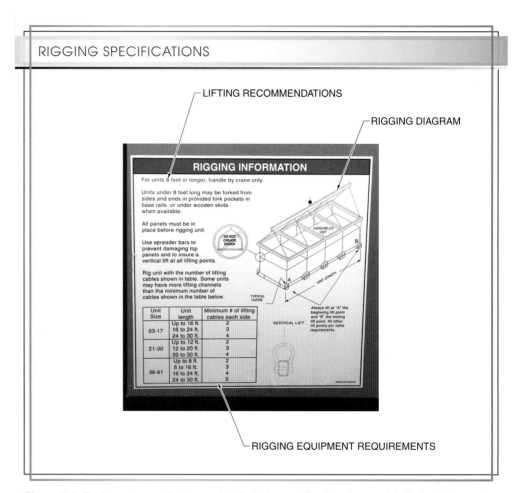

Figure 7-9. Equipment manufacturers often include specification sheets with rigging information.

It should be noted that each procedure determines the position of the CG along only one dimension of the load, either the length or the width. Each procedure can also be repeated along the other dimension to determine the two-dimensional position of the CG in the load footprint. Some of these procedures can also be used to determine the height of the CG. However, in most cases, the height of the CG is not critical. CG height is typically only important for tall loads or loads with low lift points.

Balance Procedure. The CG of a relatively small load with a flat bottom can be determined by setting it onto a small roller and noting the tipping point. **See Figure 7-10.** The roller can be a piece of pipe or cylindrical steel stock of small diameter, about 1″. One end of the load is lifted slightly, and the roller is placed under the load near the estimated position of the CG. The load is then lowered and rolled over the pivot until the load is balanced. The balancing point indicates the position of the CG along that axis.

Lift Point Procedure. The CG of a load can also be determined by measuring the weight of the load at different lift points. This procedure requires the use of a dynamometer or load cell. **See Figure 7-11.** A *dynamometer* is a device that measures linear force by measuring the rotational force (torque) that it induces on a load cell sensor. A *load cell* is a device that measures linear force with an electrical transducer. Either one of these devices can be attached in-line between a hoist and a load to measure the force on the rigging when the load is lifted.

DETERMINING CG LOCATION—BALANCE PROCEDURE

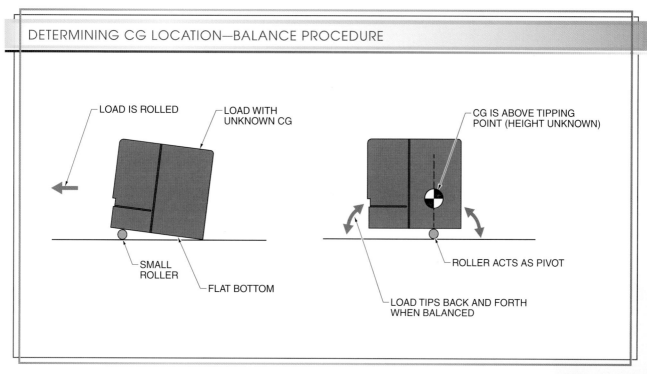

Figure 7-10. The balance procedure is a simple method for determining the location of a load's CG, but it is not appropriate for all loads.

FORCE-MEASURING DEVICES

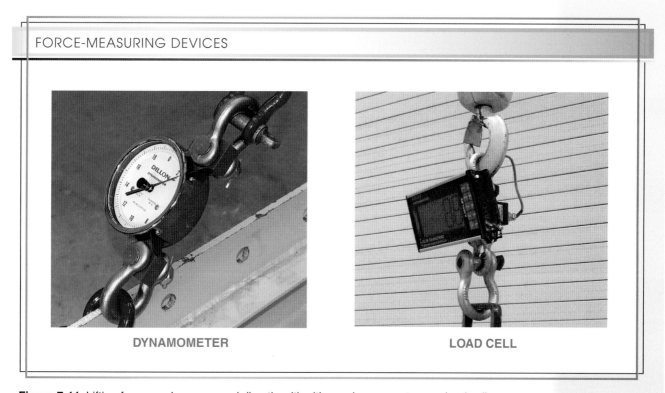

DYNAMOMETER

LOAD CELL

Figure 7-11. Lifting force can be measured directly with either a dynamometer or a load cell.

In this procedure, the measuring equipment is attached to the rigging at one end of the load. The load is then lifted slightly, and the indicated weight is recorded. **See Figure 7-12.** The measuring equipment is then moved to the opposite end of the load, and a second lift is made. The CG is always closer to the heavier end. The position of the CG along the axis between the lift points is at a distance proportional to one end's fraction of the load's total weight. This distance is calculated using the following formula:

$$d_{A-CG} = d_{A-B} \frac{W_B}{W_A + W_B}$$

where

d_{A-CG} = distance of CG along axis from point A (in ft or in.)
d_{A-B} = distance between points A and B (in ft or in.)
W_A = weight at point A (in lb)
W_B = weight at point B (in lb)

For example, a large rectangular load is slightly lifted at one end (lift point A), and a dynamometer indicates a force of 6000 lb. Then the opposite end (lift point B) is lifted, and the dynamometer indicates a force of 4000 lb. The distance between the lift points is 120″. Where is the CG along its length?

$$d_{A-CG} = d_{A-B} \frac{W_B}{W_A + W_B}$$

$$d_{A-CG} = 120 \times \frac{4000}{6000 + 4000}$$

$$d_{A-CG} = 120 \times \frac{4000}{10,000}$$

$$d_{A-CG} = 120 \times 0.4$$

$$d_{A-CG} = \mathbf{48″}$$

The CG is 48″ from the lift point at the heavy end of the load. It should be noted that this is the distance from the lift point, not necessarily the distance from the end of the load.

Section CG Procedure. A complex load may include a variety of materials or equipment in separate groups, making it nonuniform. For example, a container may be loaded with pallets of an item on one side and pallets of something heavier on the other.

If the lift point and balance procedures are not feasible, the overall CG of a complex load can be determined by the section CG procedure. This involves considering the load as a group of smaller sections, each with its own weight and CG. This method is simple but practical only if the weight and CG of each section are known to a reasonable degree. If load documentation does not contain this information, it can sometimes be estimated if a section is uniform in composition and shape. The weight can be calculated from stock material weight tables, and the section CG can be considered to be in the center of that section.

This procedure is similar to the lift point procedure, except that section weights and CG locations are used instead. Also, it can be applied to three or more sections by repeating the procedure as often as necessary until the CG for the complete load is determined.

For example, a load has three distinct sections, A, B, and C, weighing 2 t, 4 t, and 1 t, respectively. **See Figure 7-13.** The 2 t and 4 t sections are considered first. Their CGs are estimated to be in their centers, which makes them approximately 60″ apart. The combined CGs of these two sections must be closer to the CG of section B than section A since section B is heavier. The position of the combined CGs is calculated with the same proportion formula used in the lift point procedure:

DETERMINING CG LOCATION—LIFT POINT PROCEDURE

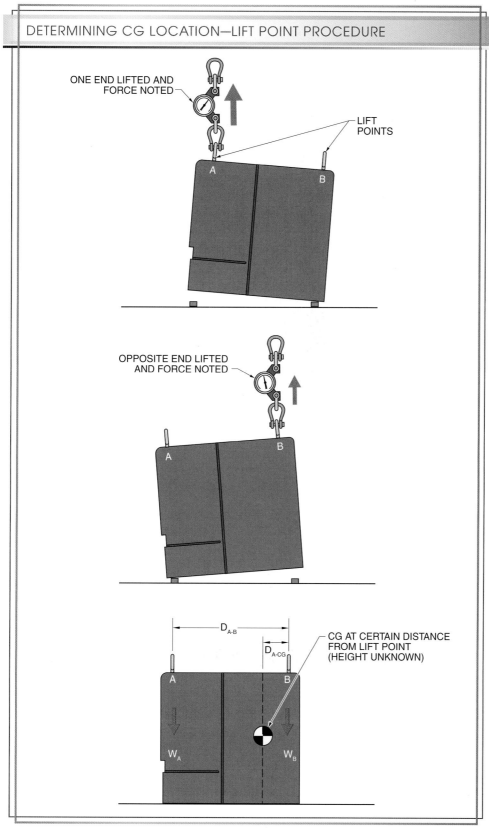

Figure 7-12. In the lift point procedure, each end of a load is lifted slightly, and the lifting force is measured. This information is then used to determine the CG location.

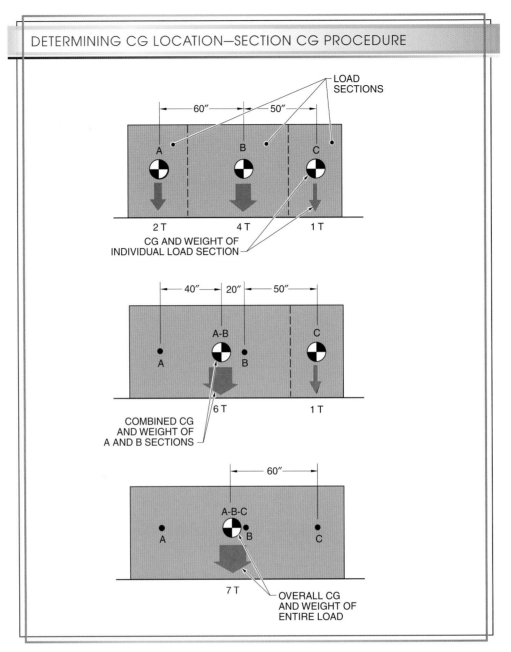

Figure 7-13. The section CG procedure involves first considering a complex load as a group of smaller sections, each with its own weight and CG. This information is then used to determine the CG of the entire load.

$$d_{A-CG} = d_{A-B} \frac{W_B}{W_A + W_B}$$

$$d_{A-CG} = 60 \times \frac{4}{2+4}$$

$$d_{A-CG} = 60 \times \frac{4}{6}$$

$$d_{A-CG} = 60 \times 0.67$$

$$d_{A-CG} = \mathbf{40''}$$

Thus, the distance between point B and the combined CGs is 20″ (60″ – 40″ = 20″). The relative section weights and orientations determine which of these distances is needed for the next calculation. A quick sketch of the load and its sections may be helpful to visualize the distances and weights. It may also be helpful to remember that a combined CG is always closer to the heavier of the two section CGs.

Next, the combined A-B CGs (from the first two sections) is compared to the CG for the third section, C, and the calculation is repeated. In this case, the final CG is shifted toward the combined A-B CG point, which has a weight of 6 t. The distance between the B and C CG is 50″. From the sketch of the load, the distance between the A-B and C CGs is determined to be 70″ (50″ + 20″ = 70″). The overall CG for the entire load is then calculated as follows:

$$d_{C-CG} = d_{C-AB} \frac{W_{AB}}{W_C + W_{AB}}$$

$$d_{C-CG} = 70 \times \frac{6}{1+6}$$

$$d_{C-CG} = 70 \times \frac{6}{7}$$

$$d_{C-CG} = 70 \times 0.86$$

$$d_{C-CG} = \mathbf{60''}$$

Therefore, the overall CG for this complex load is 60″ from the CG of section C. This is a logical result, as it is close to the center of the load but slightly closer to the 2 t section than the 1 t section.

Moments Procedure. If a load has defined sections, the overall CG for the load can also be found by calculating the moment for each section. A *moment* is the tendency of a force to rotate an object around a point. **See Figure 7-14.** For example, the heavy end of an unbalanced load tends to turn the load around a point. A moment about a point is calculated by multiplying a force by its distance from the point. Therefore, the moment for each section of a load is calculated by multiplying the section weight by the distance to the section's CG, as measured from a reference point.

Altec, Inc.

A properly balanced load should not have any moment or tipping force.

MOMENTS

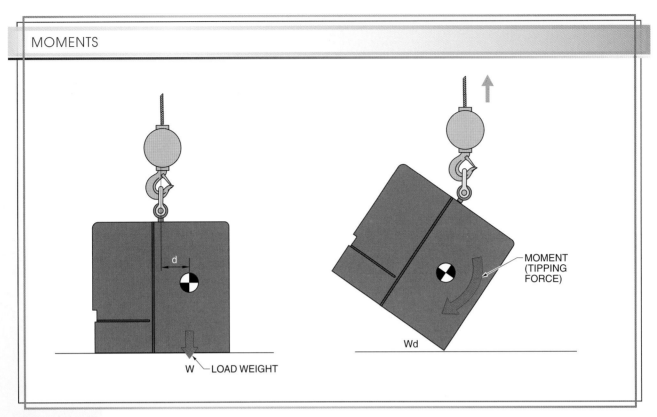

Figure 7-14. A moment is the tendency of a force, such as weight, to rotate an object around a point.

If a hoisted load is balanced, the moments that are trying to turn the load one way are counteracted by the moments trying to turn the load the opposite way. This concept can be used to determine the position of the overall CG. **See Figure 7-15.**

First, a point on the load is chosen as a reference. (Any point may be chosen, but an end or corner is most convenient.) Then, the moments induced by each section in relation to the reference point are calculated, and these are added together. This total is the same as a single equivalent moment acting at the load's overall CG. This is represented in the following formula:

$$W_{total} d_{CG} = W_A d_A + W_B d_B + W_C d_C + ...$$

where

W_{total} = total load weight (in lb or t)

d_{CG} = distance from reference point to overall CG point (in ft or in.)

W_A = weight of section A (in lb or t)

d_A = distance from reference point to CG of section A (in ft or in.)

W_B = weight of section B (in lb or t)

d_B = distance from reference point to CG of section B (in ft or in.)

W_C = weight of section C (in lb or t)

d_C = distance from reference point to CG of section C (in ft or in.)

Since the total load weight is known (the sum of all section weights), the distance to the overall CG can be calculated easily. The result is the following modification to the formula:

$$d_{CG} = \frac{W_A d_A + W_B d_B + W_C d_C + ...}{W_A + W_B + W_C + ...}$$

DETERMINING CG LOCATION—MOMENTS PROCEDURE

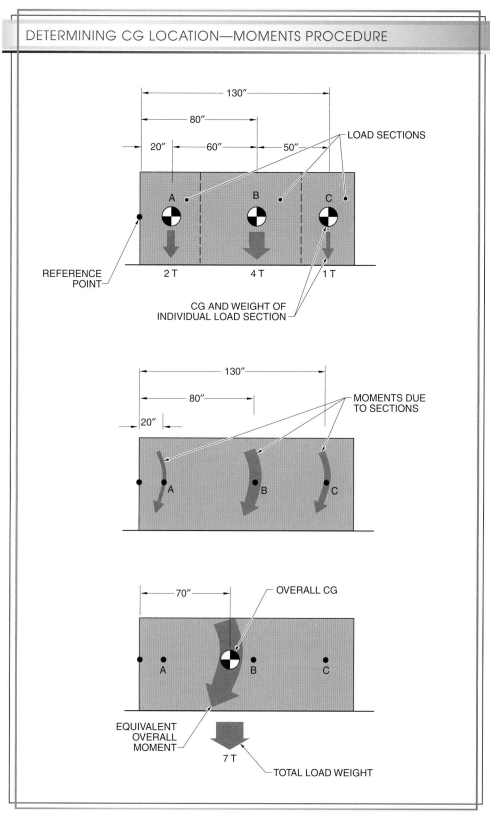

Figure 7-15. Sometimes the location of the CG of a complex load can be determined by the moments procedure, which involves adding the moments of the individual sections and dividing the sum by the total load weight.

For example, a load has three sections weighing 2 t, 4 t, and 1 t respectively. The distances of the CGs for each section from one end of the load are 20″, 80″, and 130″ respectively.

$$d_{CG} = \frac{W_A d_A + W_B d_B + W_C d_C + \dots}{W_A + W_B + W_C + \dots}$$

$$d_{CG} = \frac{2 \times 20 + 4 \times 80 + 1 \times 130}{2 + 4 + 1}$$

$$d_{CG} = \frac{40 + 320 + 130}{7}$$

$$d_{CG} = \frac{490}{7}$$

$$d_{CG} = \mathbf{70″}$$

Therefore, the CG along this dimension is located 70″ from the reference point. The position of the overall CG point in other dimensions, such as width and height, can be calculated in the same way. This is one reason why a corner is usually the most convenient reference point.

Rigging Symmetry

Once the CG of a load is determined, the load should be hoisted from directly over that point, which ensures balance. The rigging must be arranged to maintain this position and distribute the lifting force to the lift points. Since suitable lift points may be predetermined or at least restricted in options, the possible rigging arrangements are limited. The resulting rigging may be symmetrical or asymmetrical. **See Figure 7-16.** This symmetry is not necessarily related to the symmetry of the shape of the load.

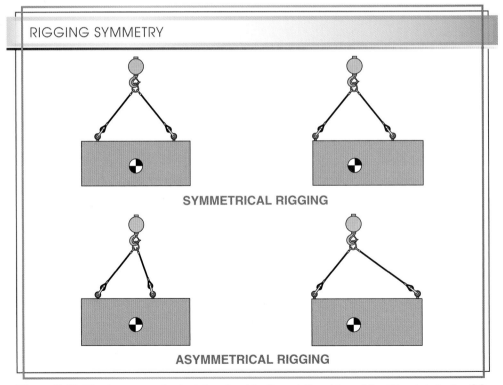

RIGGING SYMMETRY

SYMMETRICAL RIGGING

ASYMMETRICAL RIGGING

Figure 7-16. Rigging symmetry is determined by the relationships of the lift points to the CG, not necessarily by the symmetry of the load.

Also, the total load weight is divided between each lift point. For example, if a single lift point is used to hoist a load, then the entire load weight is applied to the lift point and sling. However, if a load is rigged with more than one lift point, then each point supports only a portion of the load weight, which reduces the strength requirements for the slings used. The symmetry of the rigging arrangement, along with the number of lift points used, determines how the lifting force, or the load weight, is distributed among the points.

Symmetrical Rigging. In symmetrical rigging arrangements, the lift points are all equal distances from the load's CG. This allows the use of slings that are equal in length and at equal angles.

If the rigging is symmetrical, then each lift point can be assumed to support an equal portion of the load, up to three-point arrangements. **See Figure 7-17.** In a two-point symmetrical arrangement, each point supports one-half of the load weight. For example, if symmetrical rigging of two points is used to lift a 1000-lb load, each point is supporting 500 lb. Similarly, for three-point arrangements, each point supports one-third of the load weight.

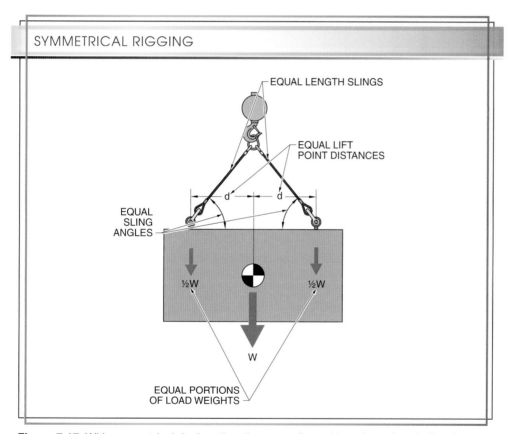

Figure 7-17. With symmetrical rigging, the slings are of equal length, and each lift point supports an equal portion of the load weight.

For symmetrical arrangements of four or more points, each point is considered to be supporting one-third of the load weight. This is because in order to equally distribute the weight, each sling would have to be so perfectly arranged in position and length that it is considered practically impossible. The fourth or any additional points are considered to be extra safety factors.

Asymmetrical Rigging. If rigging must be asymmetrical, one or more lift points are horizontally closer to the CG than the others. This means that the required sling lengths and angles are unequal. **See Figure 7-18.** The load weight is still distributed among the individual slings but not evenly. Some bear more of the load's weight than others.

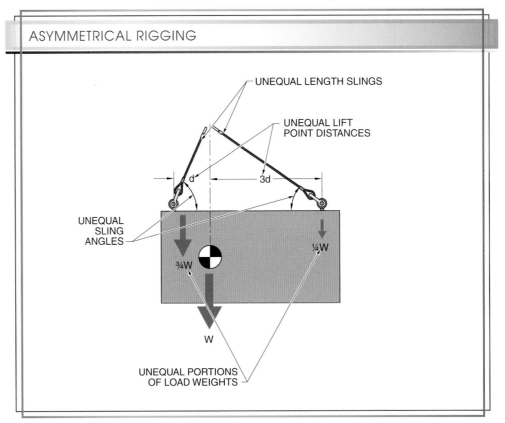

Figure 7-18. With asymmetrical rigging, the slings are of unequal length, and the lift points support unequal portions of the load weight.

In order to calculate the force experienced by each sling, the portion of load weight that is supported by each sling must be determined. This amount is proportional to the ratio of the distance from the CG to the opposite lift point to the total distance between the lift points. This is represented in the following formula:

$$W_A = W_{total} \frac{d_{CG-B}}{d_{A-B}}$$

where
W_A = portion of load weight supported at lift point A (in lb or t)
W_{total} = total load weight (in lb or t)
d_{CG-B} = horizontal distance from CG point to lift point B (in ft or in.)
d_{A-B} = horizontal distance between lift points A and B (in ft or in.)

For example, a load has lift points that are 120″ apart, but the CG is located 40″ horizontally from one of the points. **See Figure 7-19.** If the load is rigged so that it is hoisted from directly over the CG point, as it should be, the closer lift point bears a greater portion of the load's weight. If the load is 9000 lb, how is the load weight distributed between the two points?

$$W_A = W_{total}\frac{d_{CG-B}}{d_{A-B}}$$

$$W_A = 9000 \times \frac{80}{120}$$

$$W_A = 9000 \times 0.6667$$

$$W_A = \textbf{6000 lb}$$

$$W_B = W_{total}\frac{d_{CG-A}}{d_{A-B}}$$

$$W_B = 9000 \times \frac{40}{120}$$

$$W_B = 9000 \times 0.3333$$

$$W_B = \textbf{3000 lb}$$

Two-thirds of the load weight is supported at point A, the point closer to the load CG, and the remaining one-third is supported at point B.

In addition to ensuring balance, load weight and its distribution among the lift points must be calculated, or at least estimated, in order to determine the proper rigging requirements. This information is used to calculate the forces experienced by the rigging so that the proper slings and hardware can be chosen.

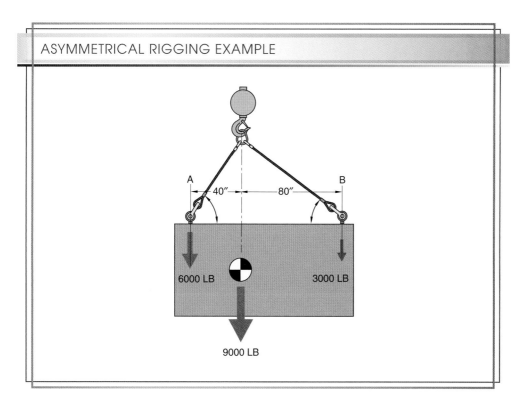

ASYMMETRICAL RIGGING EXAMPLE

Figure 7-19. The portion of load weight supported at each lift point is related to the relative distances of the lift points from the load's CG.

FORCES

Force is the interaction between two objects. Types of forces typically encountered in the outside line industry include compression, tension, shear force , and torque. **See Figure 7-20.**

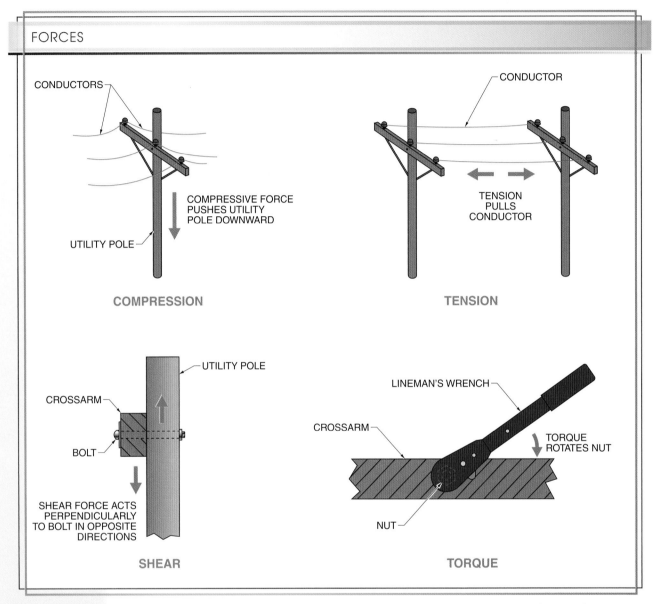

FORCES

Figure 7-20. The types of forces typically encountered in the outside line industry include compression, tension, torque, and shear.

Compression is the inward pushing force on an object. An example of compression is the weight of overhead conductors pushing downward on a utility pole. *Tension* is the outward pulling force on an object. An example of tension is the pulling of overhead conductors so that they do not sag too low. *Torque* is the rotational (turning) force on an object. An example of torque is using a wrench to tighten a nut and bolt. *Shear force* is the pushing force on one part of an object in one direction and another part of the object in the opposite direction. An example of shear force is the perpendicular forces acting on a crossarm bolt.

Another force that may be encountered is friction. *Friction* is the force of two objects or surfaces resisting movement. Friction is the force that prevents some objects from sliding down an inclined plane. Friction is also the force that causes an object in motion, sliding against another object, to eventually stop.

Work is the energy used when a force is exerted over a distance. *Energy* is the capacity to do work. *Power* is the rate of doing work or using energy.

SIMPLE MACHINES

Simple machines offer a mechanical advantage in moving loads. Simple machines include levers, wheels, pulleys, inclined planes, and screws. **See Figure 7-21.**

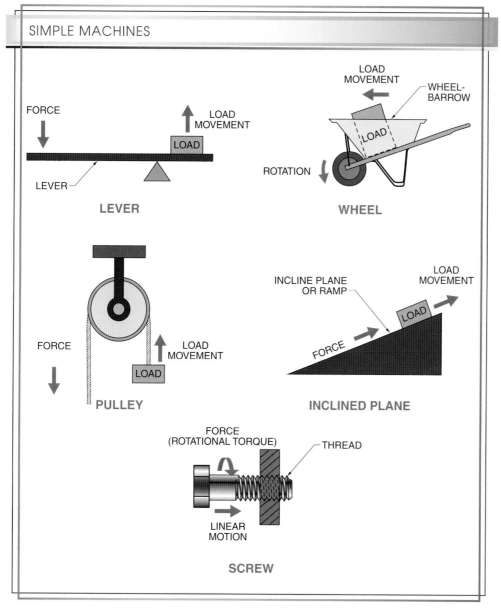

Figure 7-21. Simple machines used for mechanical advantage include levers, wheels, pulleys, inclined planes, and screws.

A *lever* is a bar that turns on a point or fulcrum. An example of a lever is a seesaw with an adjustable fulcrum to allow a small child to lift an adult.

A *wheel* is a circular object that rotates about its axis and is typically used in conjunction with an axle to easily move heavy objects. For example, the wheel on a wheelbarrow rotates and allows heavy material to be moved easily.

A *pulley* is a simple machine that consists of a grooved wheel attached to a frame or block. The grooved wheel allows rope or cable to easily pass through to change the direction of force applied to the rope or cable. Pulleys are typically referred to as blocks or block and tackle when including pulleys and rope.

An inclined plane, such as a ramp or wedge, requires less force to move a load the greater the distance that is traveled. The greater the distance the force is applied, the smaller the force required to do the work.

A *screw* is a cylindrical shaft with spiral grooves or ridges called threads. A screw converts rotational torque into linear motion.

WORKING LOAD LIMITS

The working load limit (WLL) is determined by dividing the breaking strength of a piece of rigging equipment by the safety factor. The safety factor varies depending on the application and type of material. WLL is calculated using the following formula:

$$WLL = \frac{MBS}{SF}$$

where

WLL = working load limit (in lb)

MBS = minimum breaking strength (in lb)

SF = safety factor

For example, a rope with a breaking strength of 20,000 lb and a safety factor of 5 has a WLL of 4000 lb.

$$WLL = \frac{MBS}{SF}$$

$$WLL = \frac{20,000}{5}$$

$$WLL = \textbf{4000 lb}$$

BLOCK AND TACKLE

Block and tackle is a combination of sheaves and ropes used to improve lifting efficiency. The technology was developed centuries ago to move sails, spars, and other components on sailing ships. However, this technology is still useful for hoisting, securing, and moving loads. In fact, much of the hoisting and knot terminology used today is based on a long history of nautical applications.

Blocks

A *block* is an assembly of one or more sheaves in a frame. **See Figure 7-22.** A *sheave,* also known as a pulley, is a grooved wheel attached to a frame or block that supports a rope that is changing direction.

The rest of the frame provides attachment points for either securing the block to a stationary structure or attaching the dead end of a rope. The most common types of blocks used in industrial hoisting applications are crane blocks, wire rope blocks, snatch blocks, tackle blocks, and gin blocks. **See Figure 7-23.**

BLOCKS

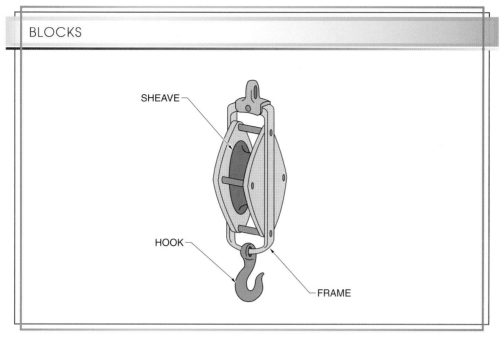

Figure 7-22. Blocks are frames containing one or more sheaves.

BLOCK TYPES

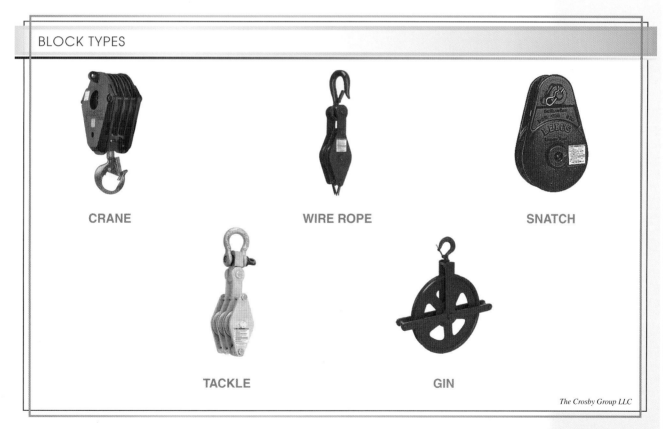

CRANE

WIRE ROPE

SNATCH

TACKLE

GIN

The Crosby Group LLC

Figure 7-23. Blocks used for rigging are determined by the number of sheaves and attached hardware required by the application.

A *crane block,* also known as a hook block, is a block used in the hook assembly of a crane. These are heavy-duty blocks that can tolerate rough handling and heavy service. A crane block is usually equipped with weights, known as cheek weights or overhaul weights, that allow it to descend smoothly via gravity when no load is applied to the block. The amount of weight needed varies according to the size of the wire rope, number of parts of rope, type of bearings used in the block, length of the boom, and amount of drum friction. Crane blocks are often equipped with lubrication fittings so the sheave bearings can be greased while in service.

A *wire rope block* is a block used with powered hoists when severe conditions of service are not expected. Two common applications of wire rope blocks are horizontal rigging and crane boom control. Wire rope blocks are relatively light and typically do not require weights to descend via gravity when being lowered.

A *snatch block,* also known as a gate block, is a block with side plates that can be opened, allowing it to be added to a hoisting line without access to the end of the line. Otherwise, the end of the line would have to be threaded through the block. Snatch blocks are available for wire rope, manila rope, and synthetic rope and can be outfitted with various combinations of hooks, eyes, shackles, or other end fittings.

A *tackle block* is a block used with natural or synthetic fiber ropes, primarily for manual hoisting operations. They are designed for slower speed operation and generally have plain bore or bronze bushing sheaves that must be disassembled for inspection and relubrication.

A *gin block,* also known as a well wheel, is a simple block that is not much more than a large sheave in a lightweight frame. These blocks are used in hand-powered hoisting operations involving relatively light loads.

Common block-and-tackle configurations include two blocks: one at the top and one at the bottom. **See Figure 7-24.** A *standing block* is the upper block in a block-and-tackle configuration that is usually attached to a fixed object or structure. A *traveling block,* also known as a fall block, is the lower moveable block in a block-and-tackle configuration that is attached to a load, usually via a hook.

Like all other rigging hardware, blocks must include identification labels that provide the specifications necessary for choosing the appropriate type and model for a task. Required information includes the manufacturer's name as well as the rated load and rope sizes intended for use with the block.

Tackle

Tackle is a combination of ropes and accessories used with blocks to gain mechanical advantage for lifting. However, the term "tackle" is rarely used on its own. When it is, it usually refers to fiber rope in nautical applications. In rigging applications, the rope is known simply as "rope" or "line."

Rope is reeved between blocks over one or more sheaves in succession. *Reeving* is the threading of a rope through an opening or around a sheave. The rope sections between blocks are known as parts. A *part* is a rope length between a dead end and block or between two blocks. One end of the rope is attached to a becket. A *becket* is an attachment point, usually on a block, for the dead end of a hoisting rope.

The remainder of the live end forms the lead line. A *lead line* is the part of a rope to which force is applied to hold or move a load. A lead line is not counted as a part.

BLOCK-AND-TACKLE MECHANICAL ADVANTAGE

The number of parts in a block-and-tackle configuration determines the mechanical advantage. *Mechanical advantage* is the ratio of the output force from a machine to the input force applied to the machine. In block-and-tackle applications, a mechanical advantage means that a load can be lifted by applying a force less than the load weight. The tradeoff is that the lifting is slower and requires a longer rope.

BLOCK AND TACKLE

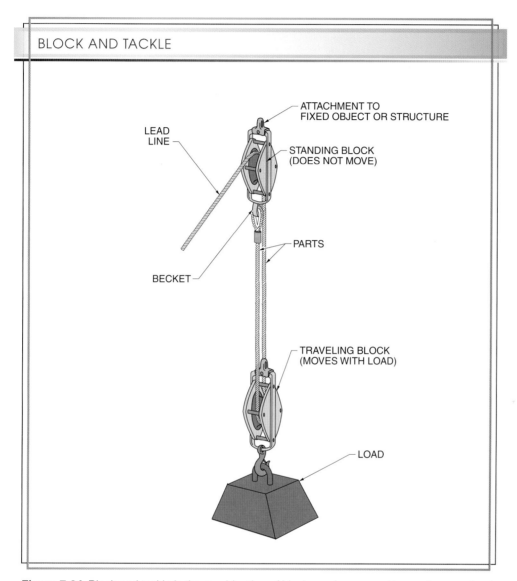

ATTACHMENT TO
FIXED OBJECT OR STRUCTURE

LEAD
LINE

STANDING BLOCK
(DOES NOT MOVE)

PARTS

BECKET

TRAVELING BLOCK
(MOVES WITH LOAD)

LOAD

Figure 7-24. Block and tackle is the combination of blocks and rope used to create a mechanical advantage.

The mechanical advantage of block and tackle is determined by its number of parts. **See Figure 7-25.** One-part reeving has one length of rope between the load and a single, upper block. There is no mechanical advantage to one-part reeving. The force required to lift the load is equal to the weight of the load.

Two-part reeving has two lengths of rope between the lower and upper blocks. The dead end of the rope is attached to the upper block, and the live end is reeved through the lower block, reeved through the upper block, and then becomes the lead line. Two-part reeving has a mechanical advantage of 2:1, reducing the lead-line force to only 50%. *Note:* The weight of the lower block must also be considered as part of the load weight, though this is usually an insignificant amount.

MECHANICAL ADVANTAGE

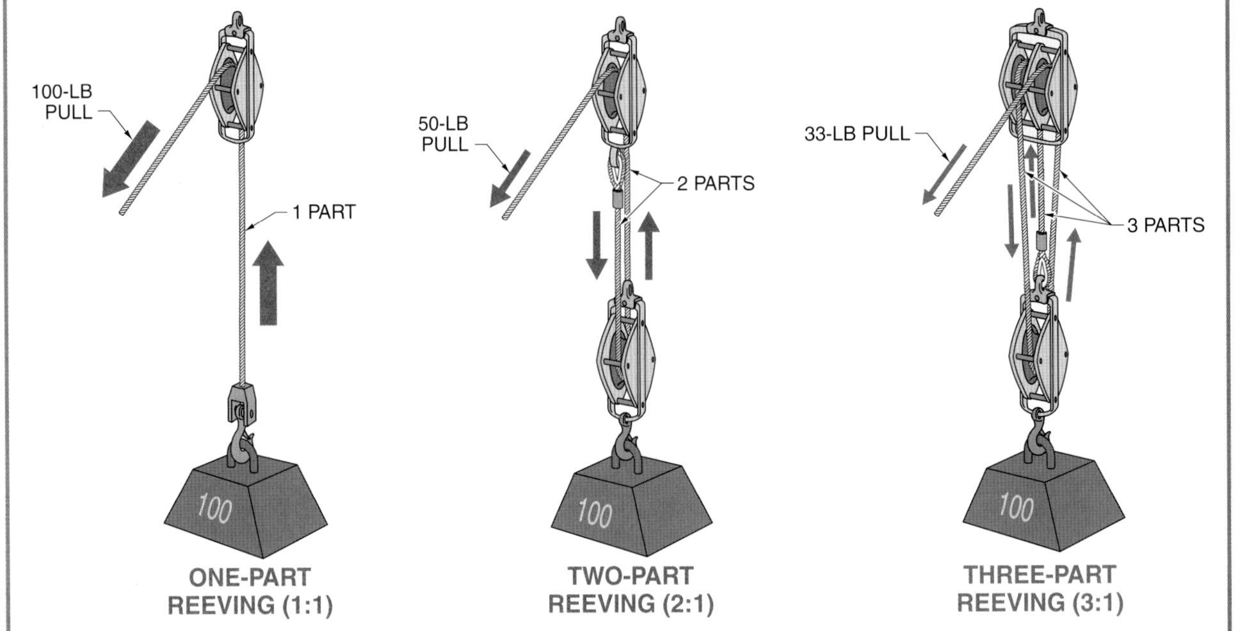

100-LB PULL

1 PART

ONE-PART REEVING (1:1)

50-LB PULL

2 PARTS

TWO-PART REEVING (2:1)

33-LB PULL

3 PARTS

THREE-PART REEVING (3:1)

Figure 7-25. Reeving rope through multiple sheaves provides mechanical advantage, meaning less force is required to lift a load.

Three-part reeving involves two sheaves in an upper block and one in a lower block. This arrangement has three parts between the load and upper block, providing a mechanical advantage of 3:1. Each part supports one-third of the load, so the force on the lead line is 33% of the load weight.

The principle of mechanical advantage using block and tackle can be extended to four or more parts in order to further improve lifting efficiency. Configurations of 10 or more parts are not uncommon, particularly in some crane types. Limitations may arise for higher-order configurations, however, due to rope length, sheave friction, and practical limitations in block design.

Forces

A study of the mechanical advantage of block and tackle requires some knowledge and analysis of the forces involved. This allows a rigger or crane operator to calculate the rated capacity necessary to lift significantly heavier loads safely and reliably.

Static Forces. A force applied to a lead line is useful only when it is equal to or greater than the static force. A *static force* is a constant force applied to an object that is only sufficient to keep the object in place. A static force is great enough to hold a load in place but not lift the load. The amount of static force required to hold a load in place is calculated using the following formula:

$$F_S = \frac{W_{total}}{n}$$

where

F_S = static lead-line force (in lb)

W_{total} = total load weight, including rigging equipment (in lb)

n = number of parts

For example, what is the force required to hold a 500 lb load using a four-part reeving system? *Note:* The rope, block, and hook components weigh a total of 30 lb.

$$F_s = \frac{W_{total}}{n}$$

$$F_s = \frac{530}{4}$$

$$F_S = \textbf{133 lb}$$

Lifting Forces. As a lead-line force exceeds a minimum static force, it overcomes friction in the sheaves, and the load begins to rise. The amount of additional force needed to begin lifting a load depends on the number of sheaves, their bend ratio, and their bearing type.

In a block-and-tackle assembly, each sheave adds friction to the system that must be overcome in order to lift a load. The number of sheaves is assumed to equal the number of parts. Bend ratio is a factor because a rope moves more easily over a larger sheave than a smaller one. The type of sheave bearing also matters because each type has different friction characteristics. **See Figure 7-26.** The axles of plain-bearing sheaves are just pins held in the frame of the block. Alternatively, ball- or roller-bearing blocks hold sheave axles with reduced-friction bearings. That is, less additional force is required to overcome the friction of roller-bearing sheaves than plain-bearing sheaves.

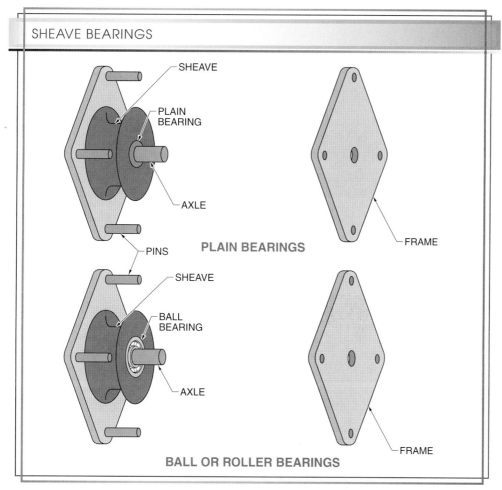

SHEAVE BEARINGS

SHEAVE
PLAIN BEARING
AXLE
PINS
FRAME
PLAIN BEARINGS

SHEAVE
BALL BEARING
AXLE
FRAME
BALL OR ROLLER BEARINGS

Figure 7-26. The type of bearing holding the sheave axle affects the amount of friction caused by rotation of the sheave.

Friction in each sheave adds a certain percentage of the load's weight as resistance. **See Figure 7-27.** Plain bearing sheaves typically add 6% to 8%. Ball- or roller-bearing sheaves typically add 3% to 5%. For example, a single sheave adding 6% in friction requires a 106-lb force to move a 100-lb load using one-part reeving.

SHEAVE FRICTION

Bend Ratio	Plain-Bearing Sheaves	Ball- or Roller-Bearing Sheaves
less than 15	8%	5%
15 to 20	7%	4%
greater than 20	6%	3%

NOTE: SHEAVE FRICTION IS TYPICALLY ROUNDED TO 10% PER SHEAVE FOR EASE OF CALCULATION

Figure 7-27. The amount of friction resistance produced by a sheave depends on the bend ratio and type of sheave bearing.

For multiple-part reeving, the load is shared by each part, but the effect of friction compounds with each sheave. In order to simplify these calculations, an appropriate friction factor is determined from a table of friction percentages and numbers of parts. **See Figure 7-28.** Then the minimum lead-line lifting force is calculated using the following formula:

$$F_L = W_{total} \times f_{fr}$$

where

F_L = lead-line lifting force (in lb)

W_{total} = total load weight (in lb)

f_{fr} = friction factor

SHEAVE FRICTION FACTORS

Number of Parts	3%	4%	5%	6%	7%	8%
1	1.03	1.04	1.05	1.06	1.07	1.08
2	0.53	0.54	0.55	0.56	0.57	0.58
3	0.36	0.37	0.39	0.40	0.41	0.42
4	0.28	0.29	0.30	0.32	0.33	0.34
5	0.23	0.24	0.26	0.27	0.28	0.29
6	0.20	0.21	0.22	0.24	0.25	0.26
7	0.18	0.19	0.20	0.21	0.23	0.24
8	0.16	0.17	0.18	0.20	0.21	0.23
9	0.14	0.16	0.17	0.19	0.20	0.22
10	0.13	0.15	0.16	0.18	0.20	0.22
11	0.13	0.14	0.16	0.17	0.19	0.21
12	0.12	0.13	0.15	0.17	0.19	0.21

Figure 7-28. Sheave friction factors are used to calculate the lead-line lifting force required to overcome sheave friction and lift a load.

For example, what is the minimum force required to lift a 6000-lb load using an eight-part reeving system equipped with plain bearing sheaves? *Note:* The bend ratio is 17. Therefore, the friction of each sheave adds an additional 7% of the load weight to the total, and the associated friction factor is 0.21.

$$F_L = W_{total} \times f_{fr}$$
$$F_L = 6000 \times 0.21$$
$$F_L = \mathbf{1260\ lb}$$

Due to the addition of friction factors, the true mechanical advantage of a block-and-tackle arrangement when lifting is less than the same arrangement when static. For example, to hold a 6000-lb load steady with an eight-part reeving system, the mechanical advantage is 8:1. The static lead-line force is only 750 lb. However, to lift the load, a minimum force of 1260 lb is required to overcome friction. The true mechanical advantage in this case is approximately 4.8:1 (6000 lb ÷ 1260 lb = 4.8).

Travel Distance

A block-and-tackle assembly amplifies force at the cost of distance. As more sheaves reduce the force required to lift a load, the distance the lead line must be pulled increases. The proportions are equal to the mechanical advantage. **See Figure 7-29.** For example, if a two-part reeve is used to lift a load by 12″, the lead line must be pulled 24″, which is a 2:1 ratio. If the assembly is a three-part reeve, then the lead line must be pulled 36″, which is a 3:1 ratio.

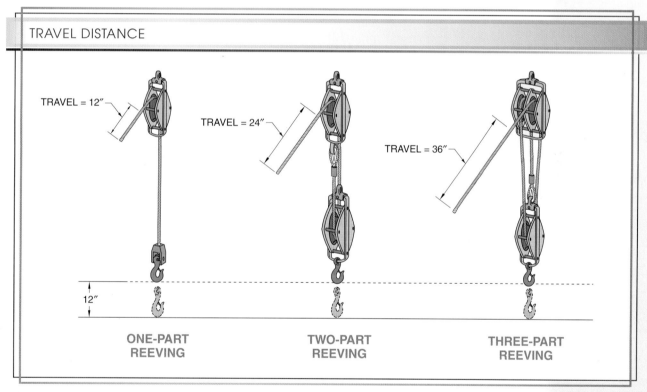

TRAVEL DISTANCE

TRAVEL = 12″

TRAVEL = 24″

TRAVEL = 36″

12″

ONE-PART
REEVING

TWO-PART
REEVING

THREE-PART
REEVING

Figure 7-29. The mechanical advantage that reduces lead-line force also increases lead-line travel. For a certain load travel distance, the lead-line travel distance must be longer.

In other words, if a load must travel a certain vertical distance, then each of the rope parts above it must shorten or lengthen by the same distance. If the travel distance is 12″ and there are three parts, then each part must change in length by 12″ for a total of 36″ of rope length. Therefore, the lead line must be pulled 36″ in order to hoist the load 12″.

Travel Speed

Travel speed is affected in the same manner as travel distance. If the lead line and the load are traveling different distances in the same amount of time, then the speeds must also be different. **See Figure 7-30.** If the lead line travels 24″ in the same amount of time that the load travels 12″, then the load speed is one-half the lead line speed. Similarly, a three-part reeved load moves at one-third the speed of the lead line.

TRAVEL SPEED

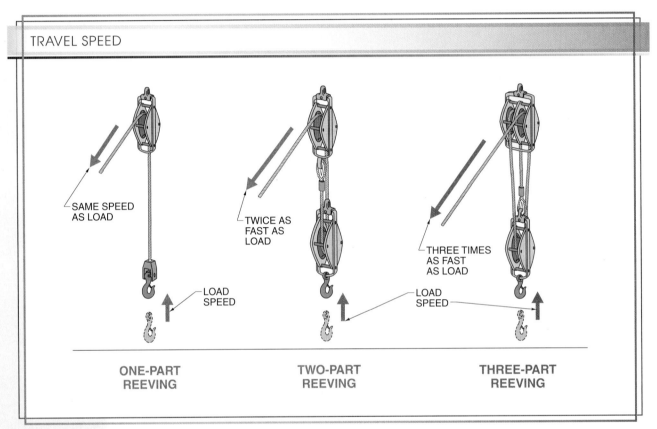

SAME SPEED AS LOAD

LOAD SPEED

TWICE AS FAST AS LOAD

LOAD SPEED

THREE TIMES AS FAST AS LOAD

LOAD SPEED

ONE-PART REEVING

TWO-PART REEVING

THREE-PART REEVING

Figure 7-30. The mechanical advantage that reduces lead-line force also increases lead-line speed. For a certain load travel speed, the lead-line travel speed must be faster.

Between the load and the lead line, each part, and therefore each sheave, is traveling at different intermediate speeds. **See Figure 7-31.** For example, in a three-part reeving, the middle sheave travels at a speed between the load and the lead line speeds. Similarly, for configurations of four parts or more, the speed of each sheave is different and gradually increases from slowest (the one nearest the dead end) to fastest (the one at the lead line). These differences may be a consideration when reeving between blocks.

REEVING

There are a number of ways to thread a rope through a set of blocks. When block-and-tackle configurations use just a few parts, the reeving method is typically not critical. However, for configurations with many parts, the differences in forces and sheave speeds make reeving an important consideration. Without distributing the load evenly, the equipment can be damaged from premature wear and slip out of place, creating a hazardous condition.

SHEAVE SPEEDS

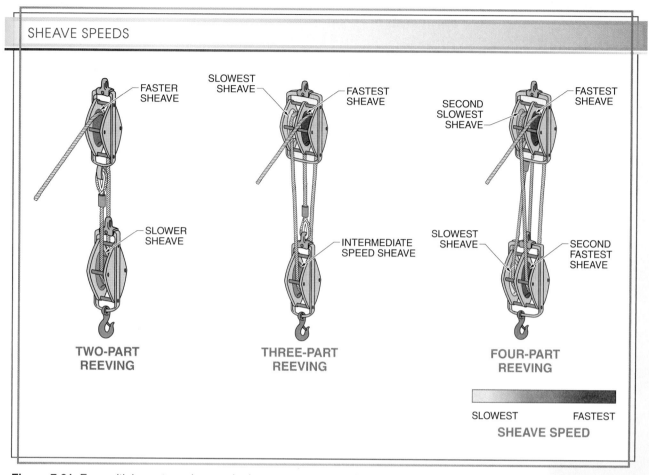

Figure 7-31. For multiple-part reeving, each sheave travels at a different speed. The sheave at the lead line is always the fastest, and the sheave nearest the load is always the slowest.

The simplest reeving method is lacing. *Lacing* is a type of reeving in which a rope starts on one side of the blocks and passes to the other side one sheave at a time. **See Figure 7-32.** The rope runs between the blocks until all of the sheaves are used, and then the dead end is attached to a becket. The rope parts on each side of the blocks are roughly parallel with each other. This method allows the blocks to be pulled close together, but the uneven force tilts the blocks to one side, which causes greater wear on the sheaves on the lead-line side.

Skip reeving is a method of reeving in which the rope starts at a center sheave and passes from one side of blocks to the other side before being attached to a becket. **See Figure 7-33.** If fewer parts are needed than sheaves available, then some sheaves can be left unreeved. Skip reeving results in more even sheave speeds across the blocks, which helps keep the blocks level with each other. However, the changes in direction at the end sheaves cause additional wear on the rope.

Note that a block load rating is based upon all of the sheaves being used. If only some of the sheaves are used, the rated load of the block is reduced proportionately. Also, the blocks with unused sheaves should be reeved symmetrically so that they do not tilt under load.

Blocks can also be reeved when at right angles to each other. *Square reeving* is a method of reeving in which a rope passes between sheaves at right angles to each other, which makes the rope change direction at each sheave. Usually, the lead line starts at one of the center sheaves, which helps to keep the blocks level, and then passes from one side to the other. **See Figure 7-34.** However, since the lines cross and change direction as they are reeved, the blocks cannot be pulled close together when in use and some long-term fatigue to the rope will occur due to the changes in direction.

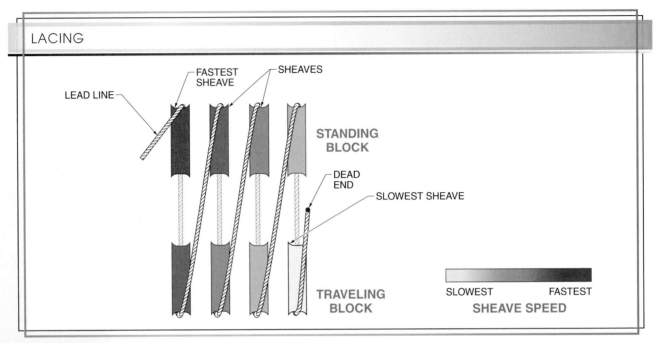

Figure 7-32. In lacing, the rope is reeved from one end of a block pair to the other. This is simple reeving, but the resulting uneven force tends to tilt the blocks.

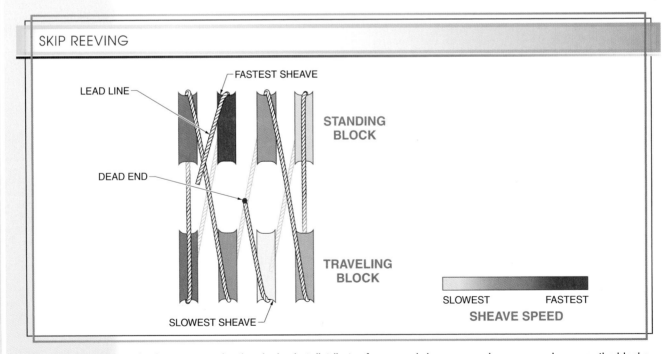

Figure 7-33. Skip reeving is more complex than lacing but distributes forces and sheave speeds more evenly across the blocks.

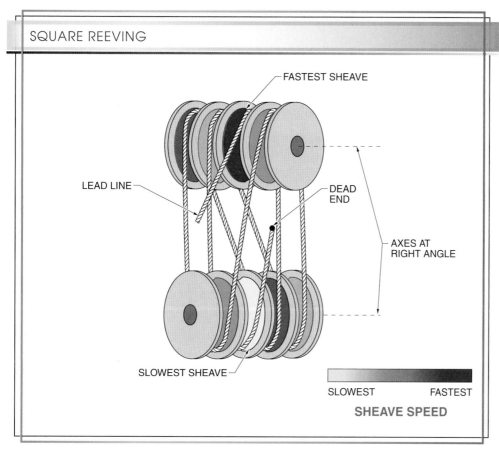

SQUARE REEVING

FASTEST SHEAVE

LEAD LINE

DEAD END

AXES AT RIGHT ANGLE

SLOWEST SHEAVE

SLOWEST   FASTEST

SHEAVE SPEED

Figure 7-34. Square reeving is a method of reeving blocks whose axes are at right angles to each other.

Other reeving methods have been developed for certain types of blocks or desirable lifting characteristics. Some can be quite complex, and the block or crane manufacturer should be consulted for correct sequences.

BLOCK LOADING

In addition to determining lifting and pulling forces, the rated load of a block must be considered. A traveling block experiences a force equal to the weight of the load. However, due to the arrangement of forces in block-and-tackle assemblies, the forces on standing blocks can exceed the load weight significantly. *Block loading* is the total amount of static force experienced by a block while in a certain arrangement. All blocks must be rated to withstand the total force they may experience.

Block Loading Estimation

If the lead line is held vertically, the standing block experiences two parallel forces, the load weight and the lead-line static force. In a single-part reeving assembly, the lead-line force is equal to the load weight. **See Figure 7-35.** For example, if the load is 100 lb, then the block experiences 200 lb of downward force. For multipart reeving, the lead-line force is less than the load weight, but the forces are added in the same way. For example, if the load is 100 lb and there are two parts, meaning that the lead-line force is 50 lb, then the block loading is 150 lb.

BLOCK LOADING

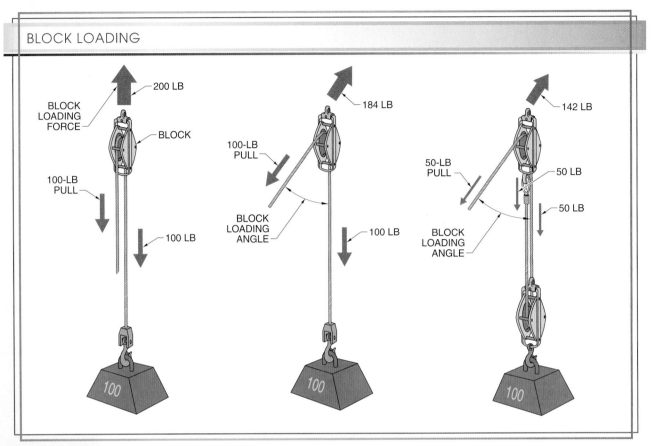

Figure 7-35. The standing block is under a greater force than just the load weight. The number of parts and the angle of lead-line affect the total block loading.

If the lead line is at an angle to the block-and-tackle assembly, the loading on the standing block is reduced somewhat and is also at an angle. This total, angular force can be calculated precisely but involves complex formulas. If the lead-line angle is small, the difference is not significant, so the block loading is often determined as if the lead line were vertical. Otherwise, the block loading can be closely estimated with a relatively simple formula that uses a block loading angle factor. **See Figure 7-36.** The total block loading is then calculated using the following formula:

$$F_{BL} = F_{LL} \times f_{BL} + (W_{total} - F_{LL})$$

where

F_{BL} = block loading (in lb)

F_{LL} = lead-line force (in lb)

f_{BL} = block loading angle factor

W_{total} = total load weight (in lb)

With this formula, a multiplier is applied to the lead-line force to account for its angle, and then the force is added due to the load weight (except for one of the parts, which is already accounted for in the block loading angle factor). For example, consider a block-and-tackle assembly holding a 1000-lb load. The assembly has four parts, so the lead-line force is 250 lb (one-quarter of the load). The lead line is pulled at a 40° angle from vertical. What is the total block loading estimate?

BLOCK LOADING ANGLE FACTORS

Angle*	Factor
0	2.00
10	1.99
20	1.97
30	1.93
40	1.87
45	1.84
50	1.81
60	1.73
70	1.64
80	1.53
90	1.41
100	1.29
110	1.15
120	1.00
130	0.84
135	0.76
140	0.68
150	0.52
160	0.35
170	0.17

* in °

Figure 7-36. The block loading angle factor is used to estimate the loading force on a standing block.

$$F_{BL} = F_{LL} \times f_{BL} + (W_{total} - F_{LL})$$
$$F_{BL} = 250 \times 1.87 + (1000 - 250)$$
$$F_{BL} = 468 + 750$$
$$F_{BL} = \mathbf{1218\ lb}$$

The actual calculated block loading, using a more complex formula, is 1202 lb. Therefore, the simplified formula generates a good estimate.

Block Loading Considerations

There are a number of considerations that arise from block loading calculation. First, the number of parts in a block-and-tackle assembly significantly affects the total block loading. If the assembly had only two parts, the block loading estimate would be 1435 lb (1419 lb actual). Therefore, the choice of block-and-tackle configuration, and not just the total load weight, affects the chance of the block loading exceeding the block's rated load.

Second, these block loading forces are static forces, which do not take into account lifting friction or dynamic (motion) forces. The increased lead-line force to overcome friction is relatively minor. However, dynamic forces due to sudden movements can be significant and can easily exceed the rated load of any of the rigging or hoisting components. This is another reason all load movements should be made slowly and changes in direction be made gradually. If these rules are followed, dynamic forces should be negligible.

Finally, block loading should be considered when determining the attachment point of the standing block. The attachment must be rated for greater than the total block loading. Also, attachments are designed primarily for vertical loads, but side loading from a lead line adds a horizontal component. The attachment method must be able to withstand both of these forces.

BLOCK INSPECTION

Blocks are manufactured for a specific size of rope, as the sheave grooves have a contour that matches the rope size. When properly matched, the sheave provides support for a rope over about a 120°-to-150° arc. If the groove is too narrow, it pinches and frays the rope. This also causes wear, known as corrugation, to the sides of the groove. If the groove is too wide, the rope flattens under load and wears the base of the groove.

The same blocks and rope are often used together, and over time, they wear together. When rope must be replaced, often the blocks must be replaced because the sheave grooves are no longer their original size. Over time, both the rope and the sheave groove change shape slightly. Sheave gauges can be used to inspect the groove of a sheave. When a gauge is inserted into a sheave, the gauge should bottom out with no gap visible at the bottom of the groove. **See Figure 7-37.** There should also be no excessive gaps at the sides of the sheave. Either condition indicates that the sheave groove is worn and must be remachined or replaced. Sheaves that are excessively worn, cracked, warped, or broken should be immediately replaced.

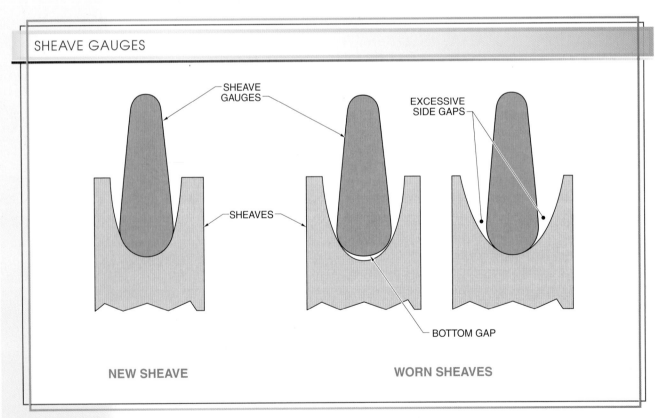

SHEAVE GAUGES

SHEAVE GAUGES

SHEAVES

EXCESSIVE SIDE GAPS

BOTTOM GAP

NEW SHEAVE

WORN SHEAVES

Figure 7-37. Sheave gauges are used to check the shape of the groove of a sheave. Worn sheaves may have gaps at the bottom or sides and should be removed from service.

Block sheaves should also be inspected for bearing wear and adequate lubrication. Any roughness or binding of the sheaves is cause for disassembly and service of the unit. This is especially important for block assemblies that have no built-in lubrication fittings, such as gin blocks.

The frames, guides, and side plates of block assemblies should be inspected daily for damage and wear. Cracking, distortion, or damage to these components is reason to remove them from service. Since hooks are often part of a block assembly, they should also be inspected using standard hook-inspection guidelines. Blocks with swivel hooks should have their swivel bearings checked for smoothness of operation and be serviced if any roughness or binding of the sheaves is detected.

COMMON APPLICATIONS—
INSTALLING POLE-MOUNTED TRANSFORMERS WITH BLOCK AND TACKLE

A pole-mounted transformer is a cylindrically shaped transformer that is installed on a utility pole and is used to transform (step up or step down) voltage. Pole-mounted transformers may need to be installed with the help of block and tackle when a digger derrick or other large lifting equipment cannot reach the work area.

A transformer gin is installed at the top of the utility pole and a block is attached to the transformer gin. An additional block is attached near the base of the utility pole to change the direction of pull. Before the transformer can be lifted, a transformer sling must be attached to the lifting points on the transformer. Then the transformer sling is attached to the hook of the block and tackle and the transformer can be hoisted into position. The transformer is secured to the utility pole with proper bolts or brackets.

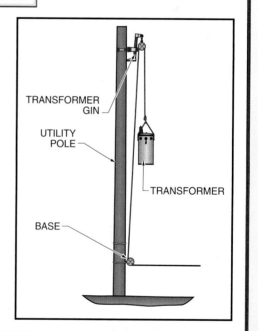

Chapter 7 Learner Resources

For additional information, visit
qr.njatcdb.org
Item # 1649

Lineworker Rigging Practices

SIGNALING 8

OBJECTIVES

- Identify the role and responsibilities of the signalperson.

- Describe methods for using multiple signalpersons during a single lift and the circumstances under which they would be needed.

- Identify the requirements for the use of nonstandard signals.

- Demonstrate proper techniques for executing both hand and voice signals.

- Compare the advantages and disadvantages of hand, voice, and audible signals.

A signalperson guides the crane operator by providing the commands needed to safely and effectively move a load during a lift. The signalperson does this by making certain signals, such as with the hands, verbally, or with sounds, that represent the desired move. Given the extreme importance of this task, signalpersons must be well-trained in proper signaling techniques and use established signals. The signalperson is also responsible for preventing hazardous situations caused by crane movements.

SIGNALPERSONS

During a lift, a crane operator may not have a clear view of a load's starting position, destination, or other points along the route of travel. This may be due to visual obstructions or the distance involved. **See Figure 8-1.** Also, other activities at a job site may need to be monitored to ensure that they will not interfere with the lift. Therefore, many hoisting operations require ground personnel to properly guide the crane operator by providing crane operating signals. A *signalperson* is a worker who is specially trained to provide clear and standardized signals to a crane operator in order to guide the movement of a load during a lift.

VISUAL OBSTRUCTIONS

Figure 8-1. Construction sites frequently include areas that the crane operator cannot clearly see from the cab. If loads must be placed in these areas, a signalperson must be used to guide the operator.

The use of a signalperson is largely determined by the conditions on the job site where the lift is to be made. Many lifts, particularly with smaller or industrial cranes, can be safely completed by the operator without the assistance of a signalperson. However, a signalperson is required if the load or any moving part of the crane will not be in full view of the operator during the entire lift. Some facilities require the use of a signalperson in all cases, even if the load will be in full view of the operator.

Signalpersons must receive appropriate training on crane movements and associated signals. Only an experienced and designated individual should act as a signalperson. In many cases, the signalperson must be certified for the task. A wrong signal could result in damage to materials or serious injury. Initial and periodic testing is used to ensure that the signalperson is qualified to perform the job correctly and safely. Signaling is also a part of the training for other lifting personnel. For example, qualified riggers are required to know how to give crane signals.

Signalperson Responsibilities

The signalperson's duty is more than just guiding the movement of a load. The signalperson must also watch for hazardous crane movements or conditions that could cause damage or injury.

The signalperson's primary duty is to provide clear and consistent signals, one at a time, to the crane operator. The signalperson must maintain constant communication with the crane operator during a lift. If at any time communication is lost, the operator must safely stop the lift and wait until communication is restored. The signalperson may even wear easily visible clothing or safety equipment, such as a yellow hard hat, in order to improve communication and ensure that no other person is mistaken as the designated signalperson. **See Figure 8-2.**

SIGNALPERSON VISIBILITY

Figure 8-2. In addition to any safety equipment required on the job site, a signalperson may also wear easily visible attire in order to improve communication with the crane operator.

A signalperson must also be familiar with the type of crane being used. All crane dynamics must be well understood so the signalperson can avoid or compensate for potential issues. For example, the crane must not be guided into a position where it may tip over or collide with another structure, which can cause significant damage.

As a load is lifted a few inches above its resting position, the lift should be paused as the rigger checks the load for stability. Depending on the circumstances, the signalperson may also hold a tag line for the load in order to keep the load from swinging or rotating into the boom or other nearby structures. This is only appropriate for relatively simple and short lifts. For more complex lifts, the signalperson concentrates on guiding the crane operator, so other personnel usually hold the tag lines.

Multiple Signalpersons

For most lifts, only one signalperson is needed. However, complex lifts may require multiple signalpersons stationed at different locations. This may be the case if a load must be guided into an enclosed space that an outside signalperson cannot see. Another signalperson may be stationed inside the space to guide the crane operator as soon as the load reaches a designated point.

Therefore, a system for switching signalpersons during a lift should be agreed upon prior to the operation. For example, a particular color or type of hard hat, vest, or gloves may be used to designate the signalperson in control. As signaling duties pass from one signalperson to another, the first signalperson removes the particular item of clothing, and the next signalperson puts on the same type of item. Or the switch in signalpersons may be clearly indicated over a radio. A variety of means may be used to indicate a switch between signalpersons as long as all personnel involved agree on the system.

The notable exception to the designated signalperson rule is that any person aware of a potential safety problem during a lift may give the Stop or Emergency Stop signal to the operator. The operator must then immediately stop the lift until the safety problem is assessed or remedied.

SIGNALS

OSHA 29 CFR 1926 Subpart CC, Cranes & Derricks in Construction, provides regulations for the safe preparation and operation of this equipment, including signaling. Specified types of crane signals include hand signals, voice signals, and audible signals. Hand signals and voice signals are the two most common types.

Nonstandard hand signals are permitted only if the standard signal set is not practical. For example, the standard signal set may not be adequate when operating special lifting equipment. In this case, nonstandard signals may be used. Whenever nonstandard signals are necessary, the signalperson and the crane operator must agree on the new signals before using them.

Hand Signals

For additional information, visit qr.njatcdb.org Item #1678

The most common method of signaling is to use hand signals. A *hand signal* is a visual signal given by a signalperson using their hands indicating a desired crane movement to a crane operator. The use of standardized hand signals is required by OSHA and ASME, so a qualified signalperson must memorize the form and meaning of all 21 standard signals. While a chart of hand signals must be posted at each job site in the vicinity of hoisting operations, a qualified signalperson should not rely on the chart to do the job.

When giving hand signals, the signalperson should ensure that their hands are fully visible to the crane operator at all times. Lighting conditions may require the signalperson to reposition themselves in order to make the signals more visible to the operator. Signaling with hands stretched away from the body or wearing brightly colored gloves may help the operator see the signals in poorly lit areas. If the sun or another light source creates a glare that interferes with the crane operator's view, the signalperson may have to turn or change position to compensate.

Stop. The Stop signal indicates that the operator should cease the crane movement and hold the load in its present position. It is a common signal used at the end of each crane movement before proceeding to the next command. Under these normal conditions, the operator will gradually slow and stop the load in order to keep the load from swinging or drifting. The Stop signal is often preceded by the Move Slowly signal so that the operator is able to anticipate the upcoming stop.

This signal is performed with one arm extended horizontally to the side. The hand is flat with palm down, and the arm is swung from front to back. **See Figure 8-3.**

Emergency Stop. The Emergency Stop signal indicates that the crane movement must be stopped immediately due to an imminent safety problem. The operator responds to this command by immediately stopping all hoisting functions. The stop will be more abrupt than a normal stop and thus may cause some swinging if the load is in motion at the time of the signal. The operator must work carefully to control the load.

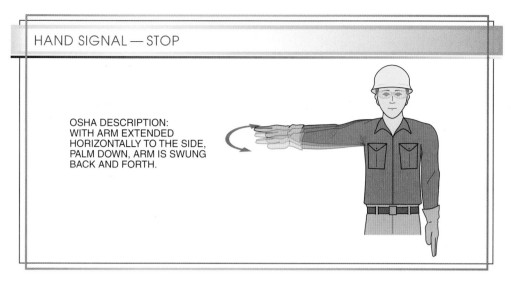

HAND SIGNAL — STOP

OSHA DESCRIPTION:
WITH ARM EXTENDED
HORIZONTALLY TO THE SIDE,
PALM DOWN, ARM IS SWUNG
BACK AND FORTH.

Figure 8-3. The Stop signal indicates that the operator should cease the crane movement and hold the load in its present position.

This signal is performed with both arms extended horizontally out from the body. The hands are flat with palms down, and the arms are swung from front to back. **See Figure 8-4.** This signal is essentially the Stop signal performed with both arms at once.

HAND SIGNAL — EMERGENCY STOP

OSHA DESCRIPTION:
WITH BOTH ARMS EXTENDED
HORIZONTALLY TO THE SIDE,
PALMS DOWN, ARMS ARE
SWUNG BACK AND FORTH.

Figure 8-4. The Emergency Stop signal indicates that the crane movement must be stopped immediately due to an imminent safety concern.

Dog Everything. The Dog Everything signal indicates that the lifting movement should be paused and that the operator should lock the crane in its present position. This pause may be needed to adjust the rigging or cribbing, reposition the suspended load, or correct some minor problem before the lift proceeds. The operator stops all movement of the load and holds the load in position when this signal is given. This signal is performed by clasping both hands together at about waist level. **See Figure 8-5.**

Use Main Hoist. Many cranes have more than one hoist. With these cranes, the hoist with the higher load rating is designated as the main hoist. The Use Main Hoist signal indicates that the main hoist should be used for the following commands.

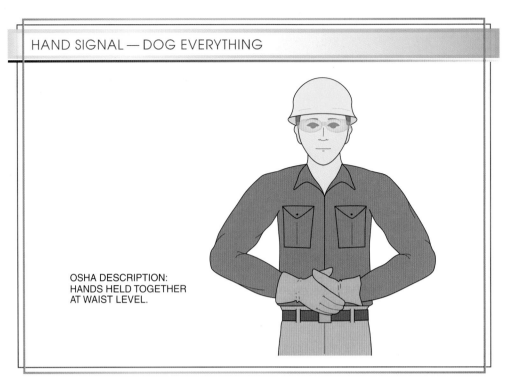

HAND SIGNAL — DOG EVERYTHING

OSHA DESCRIPTION:
HANDS HELD TOGETHER
AT WAIST LEVEL.

Figure 8-5. The Dog Everything signal indicates that the lifting movement should be paused and that the operator should lock the crane in its present position.

This signal is performed by tapping a closed fist on the top of the head. **See Figure 8-6.** Then additional crane signals follow. The operator will use the main hoist for all the following crane commands until directed otherwise.

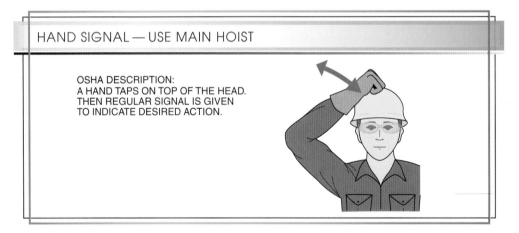

HAND SIGNAL — USE MAIN HOIST

OSHA DESCRIPTION:
A HAND TAPS ON TOP OF THE HEAD.
THEN REGULAR SIGNAL IS GIVEN
TO INDICATE DESIRED ACTION.

Figure 8-6. The Use Main Hoist signal indicates that the main hoist should be used for the following commands.

Use Auxiliary Hoist. With cranes having more than one hoist, the auxiliary hoist, or "whip line," is a lower capacity hoist used to lift lighter loads. The Use Auxiliary Hoist signal indicates that the auxiliary hoist should be used for the following commands. When using the auxiliary hoist, the rigger, operator, and signalperson should compensate for the reduced capacity and limit the load accordingly.

This signal is performed by tapping an open hand on the elbow of the other arm, which is bent with the forearm upright. **See Figure 8-7.** Then additional crane signals follow. The operator will use the auxiliary hoist for all the following crane commands until directed otherwise.

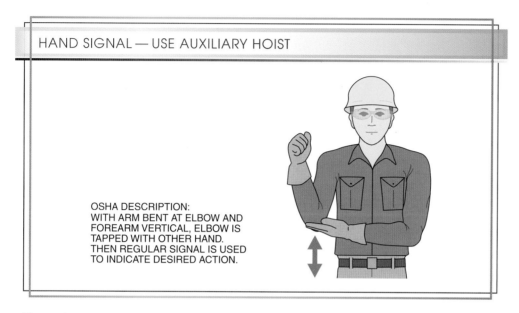

HAND SIGNAL — USE AUXILIARY HOIST

OSHA DESCRIPTION:
WITH ARM BENT AT ELBOW AND
FOREARM VERTICAL, ELBOW IS
TAPPED WITH OTHER HAND.
THEN REGULAR SIGNAL IS USED
TO INDICATE DESIRED ACTION.

Figure 8-7. The Use Auxiliary Hoist Signal indicates that the auxiliary hoist should be used for the following commands.

Hoist. The Hoist signal indicates that the load should be raised using the crane's hoist drum. As the cable retracts, the block and the hook rise, lifting the load. Care should be taken that the weight of the load does not cause excessive deflection of the boom. This signal is performed by pointing up with an index finger and moving the hand in small circles. **See Figure 8-8.**

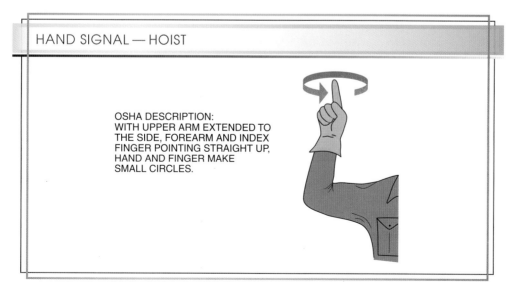

HAND SIGNAL — HOIST

OSHA DESCRIPTION:
WITH UPPER ARM EXTENDED TO
THE SIDE, FOREARM AND INDEX
FINGER POINTING STRAIGHT UP,
HAND AND FINGER MAKE
SMALL CIRCLES.

Figure 8-8. The Hoist signal indicates that the load should be raised using the crane's hoist drum.

Move Slowly. The Move Slowly signal indicates that the movement indicated by the next signal or set of signals should be performed slowly. It can be used with any signal but is most commonly used with the Hoist and Lower signals.

This signal is performed by holding the flat palm of one hand over the other hand as it gives the action signal. **See Figure 8-9.** Most demonstrations of this signal use it in conjunction with the Hoist signal.

HAND SIGNAL — MOVE SLOWLY

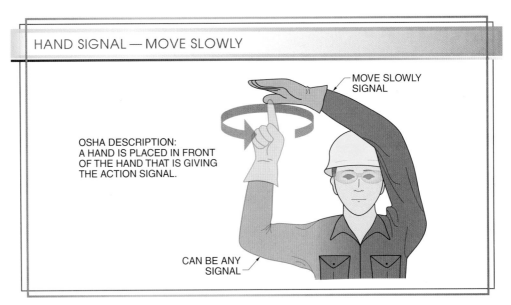

MOVE SLOWLY SIGNAL

OSHA DESCRIPTION:
A HAND IS PLACED IN FRONT OF THE HAND THAT IS GIVING THE ACTION SIGNAL.

CAN BE ANY SIGNAL

Figure 8-9. The Move Slowly signal indicates that the movement indicated should be performed slowly.

Lower. The Lower signal indicates that the load should be lowered using the crane's hoist drum. This signal is the opposite of the Hoist signal. As the load is landed, tension is released from the hoisting line, and the boom straightens slightly. It may be necessary to lower the boom slightly in order to maintain proper alignment between the boom sheave and the load. This signal is performed by pointing down with an index finger and moving the hand in small circles. **See Figure 8-10.**

HAND SIGNAL — LOWER

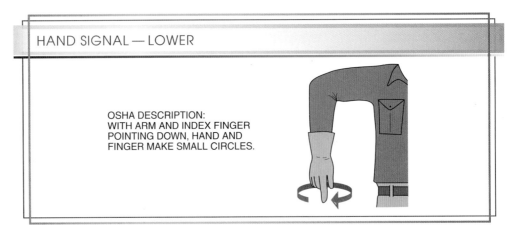

OSHA DESCRIPTION:
WITH ARM AND INDEX FINGER POINTING DOWN, HAND AND FINGER MAKE SMALL CIRCLES.

Figure 8-10. The Lower signal indicates that the load should be lowered using the crane's hoist drum.

Swing. The Swing signal indicates that the upperworks of the crane should be rotated in the direction indicated. The crane moves the load in an arc while maintaining the load's distance from the pivot point. Care should be taken that the boom is not subject to side loading from the crane moving too quickly. This signal is performed by extending one arm and pointing with an index finger in the desired swing direction. **See Figure 8-11.**

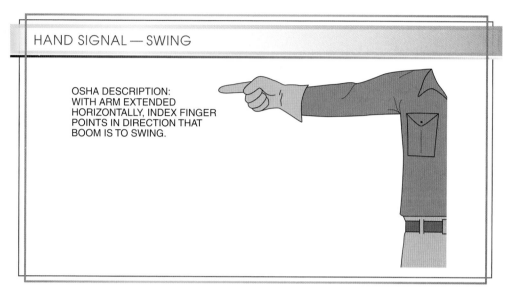

HAND SIGNAL — SWING

OSHA DESCRIPTION:
WITH ARM EXTENDED
HORIZONTALLY, INDEX FINGER
POINTS IN DIRECTION THAT
BOOM IS TO SWING.

Figure 8-11. The Swing signal indicates that the upperworks of the crane should be rotated in the direction indicated.

Raise Boom. The Raise Boom signal indicates that the boom angle should be increased, which raises the far end of the boom. The load is simultaneously moved toward the crane and up from the ground. Raising the boom decreases the load radius, which increases the load capacity of the crane.

This signal is performed by pointing the thumb of one hand upward with the arm extended. The fist remains closed. **See Figure 8-12.**

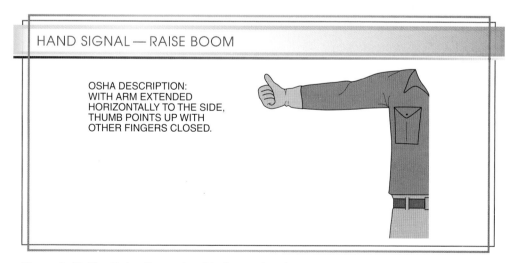

HAND SIGNAL — RAISE BOOM

OSHA DESCRIPTION:
WITH ARM EXTENDED
HORIZONTALLY TO THE SIDE,
THUMB POINTS UP WITH
OTHER FINGERS CLOSED.

Figure 8-12. The Raise Boom signal indicates that the boom angle should be increased, raising the load.

Raise the Boom and Lower the Load. The Raise the Boom and Lower the Load signal indicates that these two movements should be performed simultaneously. This is done if the load must remain at a constant elevation as the boom is raised. These movements maintain the elevation of the load as it moves horizontally closer to the crane.

This signal is performed by pointing the thumb of one hand upward with the arm extended. The fingers of this hand open and close for as long as the movement is desired. **See Figure 8-13.**

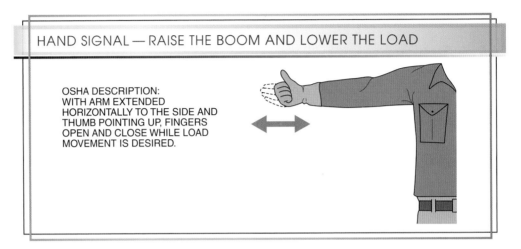

HAND SIGNAL — RAISE THE BOOM AND LOWER THE LOAD

OSHA DESCRIPTION:
WITH ARM EXTENDED
HORIZONTALLY TO THE SIDE AND
THUMB POINTING UP, FINGERS
OPEN AND CLOSE WHILE LOAD
MOVEMENT IS DESIRED.

Figure 8-13. The Raise the Boom and Lower the Load signal indicates that these two movements should be performed simultaneously.

Lower Boom. The Lower Boom signal indicates that the boom angle should be decreased, which lowers the far end of the boom. The load is simultaneously moved away from the crane and down to the ground. Caution should be used when lowering the boom because the increased load radius reduces the crane's load capacity.

This signal is performed by pointing the thumb of one hand downward with the arm extended. The fist remains closed. **See Figure 8-14.**

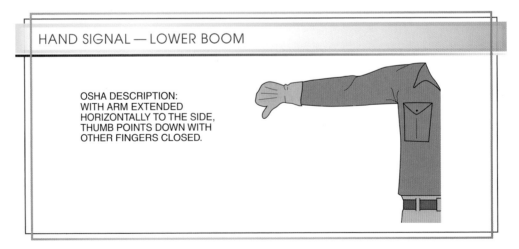

HAND SIGNAL — LOWER BOOM

OSHA DESCRIPTION:
WITH ARM EXTENDED
HORIZONTALLY TO THE SIDE,
THUMB POINTS DOWN WITH
OTHER FINGERS CLOSED.

Figure 8-14. The Lower Boom signal indicates that the boom angle should be decreased, lowering the load.

Lower the Boom and Raise the Load. The Lower the Boom and Raise the Load signal indicates that these two movements should be performed simultaneously. This is done if the load must remain at a constant elevation as the boom is lowered. These movements maintain the elevation of the load as it moves horizontally away from the crane.

This signal is performed by pointing the thumb of one hand downward with the arm extended. The fingers of this hand open and close for as long as the movement is desired. **See Figure 8-15.**

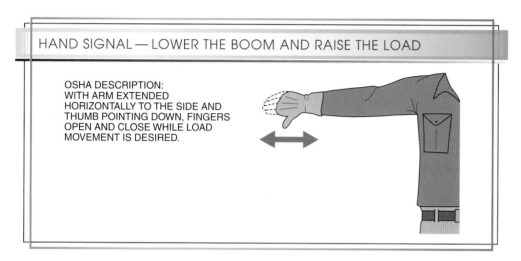

HAND SIGNAL — LOWER THE BOOM AND RAISE THE LOAD

OSHA DESCRIPTION:
WITH ARM EXTENDED
HORIZONTALLY TO THE SIDE AND
THUMB POINTING DOWN, FINGERS
OPEN AND CLOSE WHILE LOAD
MOVEMENT IS DESIRED.

Figure 8-15. The Lower the Boom and Raise the Load signal indicates that these two movements should be performed simultaneously.

Extend Telescoping Boom. The Extend Telescoping Boom signal indicates that a telescoping boom should be lengthened. Extending the boom moves the load away from the crane, increasing the load radius. This reduces the crane's load capacity and increases the chance of tipping.

This movement also causes the load to rise as the hoist line stays the same length. The signalperson must watch the crane for two-blocking, which is when the bottom block or hook is drawn up into the sheave on the top of the boom as the boom is extended. If the signalperson observes that two-blocking is about to occur, the Stop signal should be given to stop the Extend Telescoping Boom movement. Then the Lower signal should be given to provide more line between the block and the boom. The Extend Telescoping Boom signal is performed by holding both fists in front of the body at waist level and pointing the thumbs to the sides. **See Figure 8-16.**

There is also a commonly accepted one-handed version of the Extend Telescoping Boom signal that can be used if the signalperson has only one hand free. However, since this is not an OSHA-approved standard hand signal, it should be used only after the operator and the signalperson have agreed upon its meaning. The one-handed version of this signal is performed by holding the free hand in a fist in front of the body and tapping the chest with the tip of the extended thumb. **See Figure 8-17.**

Retract Telescoping Boom. The Retract Telescoping Boom signal indicates that a telescoping boom should be shortened. If it is not feasible to move the load closer to the crane by raising the boom, the same thing can be accomplished by retracting the boom if possible.

HAND SIGNAL — EXTEND TELESCOPING BOOM

OSHA DESCRIPTION:
WITH HANDS TO THE FRONT AT
WAIST LEVEL, THUMBS POINT
OUTWARD WITH OTHER
FINGERS CLOSED.

Figure 8-16. The Extend Telescoping Boom signal indicates that the telescoping boom should be lengthened.

HAND SIGNAL — EXTEND TELESCOPING BOOM (ONE-HANDED)

FREE HAND IN A FIST IN
FRONT OF BODY AND
TAPPING CHEST WITH
EXTENDED THUMB.

NOTE: NOT AN OSHA
STANDARD SIGNAL

Figure 8-17. A common alternative to the Extend Telescoping Boom signal requires only one hand.

This movement results in a lower vertical height of the crane and decreases the load radius, which increases the crane's load rating. As the boom is shortened, the load also moves downward, which may require the signalperson to give the Hoist signal in order to maintain the height of the load. The Retract Telescoping Boom signal is performed by holding both fists in front of the body at waist level and pointing the thumbs at each other. **See Figure 8-18.**

HAND SIGNAL — RETRACT TELESCOPING BOOM

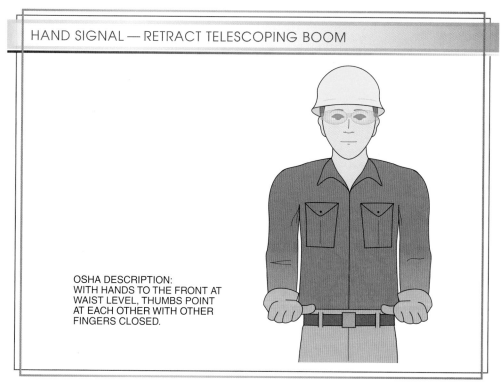

OSHA DESCRIPTION:
WITH HANDS TO THE FRONT AT
WAIST LEVEL, THUMBS POINT
AT EACH OTHER WITH OTHER
FINGERS CLOSED.

Figure 8-18. The Retract Telescoping Boom signal indicates that the telescoping boom should be shortened.

This signal also has a one-handed version, which can be used under the same conditions as its corresponding one-handed Extend Telescoping Boom signal. The one-handed version of the Retract Telescoping Boom signal is performed by holding the free hand in a fist with thumb extended in front of the body and tapping the chest with the heel of the fist. **See Figure 8-19.**

HAND SIGNAL — RETRACT TELESCOPING BOOM (ONE-HANDED)

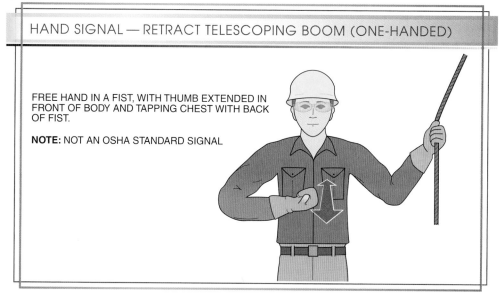

FREE HAND IN A FIST, WITH THUMB EXTENDED IN
FRONT OF BODY AND TAPPING CHEST WITH BACK
OF FIST.

NOTE: NOT AN OSHA STANDARD SIGNAL

Figure 8-19. A common alternative to the Retract Telescoping Boom signal requires only one hand.

Travel/Tower Travel. The Travel/Tower Travel signal indicates that the crane should be moved laterally in the indicated direction. This signal may be used for industrial and sometimes tower cranes. These cranes travel on rails to allow lateral motion along the length of the rails. This signal may also be used with mobile cranes. These cranes are moved either before a lift in order get into the best position to hoist a load or during a lift in order to transport the load a greater distance than possible with the boom alone. However, it should be noted that a moving mobile crane has a lower load capacity because it cannot use its outriggers. This signal is performed by extending one arm forward with palm open and fingers pointing up and making a repeated pushing motion in the direction of desired travel. **See Figure 8-20.**

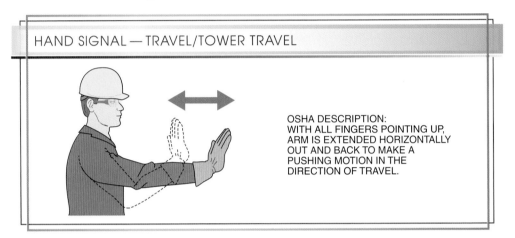

HAND SIGNAL — TRAVEL/TOWER TRAVEL

OSHA DESCRIPTION:
WITH ALL FINGERS POINTING UP, ARM IS EXTENDED HORIZONTALLY OUT AND BACK TO MAKE A PUSHING MOTION IN THE DIRECTION OF TRAVEL.

Figure 8-20. The Travel/Tower Travel signal indicates that the crane should be moved laterally in the indicated direction.

Trolley Travel. The Trolley Travel signal indicates that the hoist trolley should be moved laterally in the indicated direction. This signal is only used with cranes that have movable trollies, such as overhead, jib, gantry, and tower cranes. In these cases, the trolley travels along the boom in a direction that is typically at a right angle to the direction of travel possible by the rest of the crane. This signal is performed by extending one arm up with fist closed and thumb extended and moving the hand back and forth horizontally in the direction of desired travel. **See Figure 8-21.**

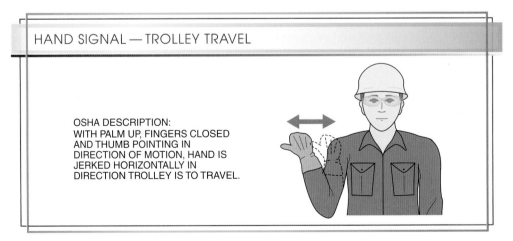

HAND SIGNAL — TROLLEY TRAVEL

OSHA DESCRIPTION:
WITH PALM UP, FINGERS CLOSED AND THUMB POINTING IN DIRECTION OF MOTION, HAND IS JERKED HORIZONTALLY IN DIRECTION TROLLEY IS TO TRAVEL.

Figure 8-21. The Trolley Travel signal indicates that the hoist trolley should be moved laterally in the indicated direction.

Crawler Crane Travel, Both Tracks. The Crawler Crane Travel, Both Tracks signal indicates that both tracks of a crawler crane should be used to move the crane straight forward or backward. This signal is only used with cranes equipped with crawler tracks that move the entire crane.

This signal is performed with both fists rotating around each other in a circular motion in front of the body. Rotating the fists in a direction away from the body indicates that the crawler crane should use the crawler tracks to move forward. Rotating the fists in a direction toward the body indicates that the crawler crane should use the crawler tracks to move backward. **See Figure 8-22.**

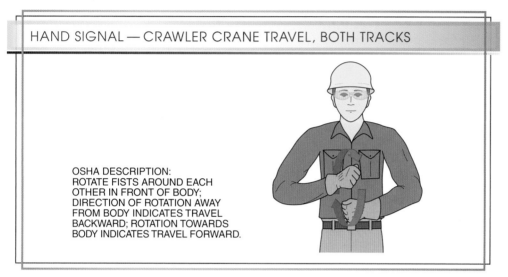

HAND SIGNAL — CRAWLER CRANE TRAVEL, BOTH TRACKS

OSHA DESCRIPTION:
ROTATE FISTS AROUND EACH OTHER IN FRONT OF BODY; DIRECTION OF ROTATION AWAY FROM BODY INDICATES TRAVEL BACKWARD; ROTATION TOWARDS BODY INDICATES TRAVEL FORWARD.

Figure 8-22. The Crawler Crane Travel, Both Tracks signal indicates that both tracks of a crawler crane should be used to move the crane straight forward or backward.

It is important to note that this signal always indicates movement from the crane operator's perspective. Therefore, the hand motions should mimic the motion of the tracks of the crane.

Crawler Crane Travel, One Track. The Crawler Crane Travel, One Track signal indicates that only one track of a crawler crane should be rotated while the other is locked in place. This action turns the crane. This signal conveys which track is to be kept motionless and which is to be rotated, along with the direction of rotation. Again, the signal is given from the operator's perspective.

This signal is performed by holding one fist in the air, indicating the track side to be held stationary, with the other fist rotating in a circular motion in front of the chest. The direction of rotation indicates the direction of desired travel. **See Figure 8-23.**

Erickson Incorporated
Hand signals or voice signals are typically used when using helicopter cranes to lift a load.

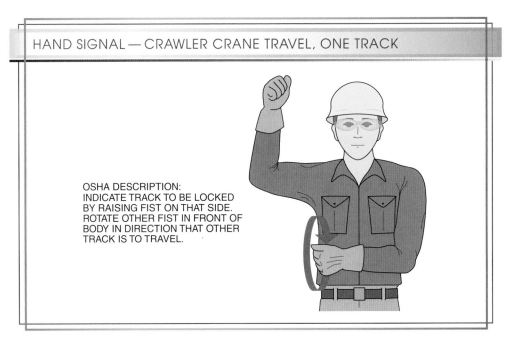

HAND SIGNAL — CRAWLER CRANE TRAVEL, ONE TRACK

OSHA DESCRIPTION:
INDICATE TRACK TO BE LOCKED
BY RAISING FIST ON THAT SIDE.
ROTATE OTHER FIST IN FRONT OF
BODY IN DIRECTION THAT OTHER
TRACK IS TO TRAVEL.

Figure 8-23. The Crawler Crane Travel, One Track signal indicates that only one track of a crawler crane should be rotated while the other is locked in place.

Helicopter Hand Signals

Helicopter cranes are specifically designed to lift heavy loads with a long hoist or sling. Helicopter cranes are typically used to lift loads in remote and inaccessible areas where it is not feasible to use other types of cranes. Helicopter hand signals have been specifically developed for use with helicopter crane lifts. **See Figure 8-24.**

Voice Signals

Voice signals are used in cases where the operator cannot see the signalperson for some or all of the lift. A *voice signal* is a signal given vocally by a signalperson indicating a desired crane movement to a crane operator. Prior to beginning a lift, the operator and signalperson must agree upon the voice signals and communication system to be used.

Voice signals, be it via radio, telephone, or another method, must be transmitted through a dedicated channel with no interference from other sources or users. **See Figure 8-25.** The crane operator must be able to receive the signals via a hands-free system, which ensures that both hands are available to manipulate the controls of the crane anytime during the lift. All communication equipment should be checked for proper operation prior to the lift, and extra batteries should be kept on hand for those devices that require them.

Prior to the lift, the signalperson and the operator should identify each other by name to ensure that the communication equipment is working and that they are on the correct channel. Constant communication must be maintained during the lift. If at any time during the lift the operator identifies a communication or hoisting problem, the lift must be stopped, and the operator should sound the crane's horn in order to alert the signalperson. Any problems must be remedied before the lift resumes.

Multiple cranes may share a common communication channel, but the signalperson and operators must designate each crane with a unique name or title. When giving voice signals to multiple cranes, the signalperson must indicate which crane each signal is for prior to giving the signal.

HELICOPTER HAND SIGNALS...

MOVE RIGHT

OSHA DESCRIPTION: LEFT ARM EXTENDED HORIZONTALLY; RIGHT ARM SWEEPS UPWARD TO POSITION OVER HEAD

HOLD HOVER

OSHA DESCRIPTION: "HOLD" IS EXECUTED BY PLACING ARMS OVER HEAD WITH CLENCHED FISTS

MOVE LEFT

OSHA DESCRIPTION: RIGHT ARM EXTENDED HORIZONTALLY; LEFT ARM SWEEPS UPWARD TO POSITION OVER HEAD

TAKE OFF

OSHA DESCRIPTION: RIGHT HAND BEHIND BACK; LEFT HAND POINTING UP

MOVE FORWARD

OSHA DESCRIPTION: COMBINATION OF ARM AND HAND MOVEMENT IN A COLLECTING MOTION PULLING TOWARD BODY

LAND

OSHA DESCRIPTION: ARMS CROSSED IN FRONT OF BODY AND POINTING DOWNWARD

Figure 8-24. Helicopter hand signals have been specifically developed for use with helicopter crane lifts.

...HELICOPTER HAND SIGNALS

MOVE REARWARD

OSHA DESCRIPTION: HANDS ABOVE ARM, PALMS OUT USING A NOTICEABLE SHOVING MOTION

MOVE UPWARD

OSHA DESCRIPTION: ARMS EXTENDED, PALMS UP; ARMS SWEEPING UP

RELEASE SLING LOAD

OSHA DESCRIPTION: LEFT ARM HELD DOWN AWAY FROM BODY; RIGHT ARM CUTS ACROSS LEFT ARM IN A SLASHING MOVEMENT FROM ABOVE

MOVE DOWNWARD

OSHA DESCRIPTION: ARMS EXTENDED, PALMS DOWN; ARMS SWEEPING DOWN

Figure 8-24. ...continued

TWO-WAY RADIO COMMUNICATION

MICROPHONE

RADIO

Figure 8-25. Two-way radios are typically used for giving voice signals to a crane operator.

The purpose of giving voice signals is to give the operator an accurate understanding of what is to be done using the appropriate sequence of commands. Clear, consistent, and continuous voice signals are necessary to safely complete each task. Each voice signal consists of three elements, which are given in the following specific order:

1. function and direction
2. distance remaining and/or speed
3. function and stop

The function is the name of the crane movement. The name used should match the name defined for the function's hand signal. For example, "hoist" should be used, not "raise the load." Also, if a direction is applicable, such as "swing right," it is given as if from the operator's perspective. That is, the boom is to swing to the operator's right, which may be the signalperson's left.

The distance remaining and/or speed information is conveyed throughout the operation in order to provide continual feedback to the crane operator. Distances should always be given as the distance remaining, not the distance that has already been traveled.

To stop the movement, the function is repeated and immediately followed by the Stop signal. For example, when a crane completes a movement, it is correct to say "hoist stop" or "swing stop," but not "stop hoist" or "stop swing."

Examples of proper voice signals include the following:

- "Hoist, 50 feet…40 feet…30 feet…20 feet…10 feet…slow…slow…hoist stop."
- "Hoist slow…slow…hoist about 20 feet…15 feet…10 feet…5 feet…slow…hoist stop."
- "Extend telescoping boom…25 feet…20 feet…15 feet…10 feet…slow…slow… 4 feet…2 feet…extend telescoping boom stop."
- "Swing left slow…15 feet…10 feet…5 feet…slow…slow…1 foot…swing stop."
- "Boom down…10 feet…5 feet…slow…2 feet…1 foot…boom down stop."
- "Lower…15 feet…10 feet…slow…5 feet…3…2…1…lower stop."

Audible Signals

Audible signals are uncommon but may be necessary in certain circumstances. An *audible signal* is a signal given by a signalperson with a sound-making device indicating a desired crane movement to a crane operator. For example, a horn, bell, or siren could be used to give audible signals. These types of signals may be useful in extremely noisy environments or areas with significant radio interference, but only if the necessary crane movements are very simple.

Audible signals can only indicate a limited set of instructions, usually those related to the most basic operation of a crane. For example, one, two, and three short sounds can indicate Stop, Forward, and Backward respectively. The operator and signalperson must meet and agree upon the signaling system prior to a lift.

COMMON APPLICATIONS—
SETTING UTILITY POLES

A utility pole is an object or column that is used to support overhead power lines. Utility poles may be made of wood, metal, fiberglass, or concrete. Utility poles are typically installed using a digger derrick.

A digger derrick is used to make the hole where the utility pole will be placed and is used to lift the utility pole into place. The digger derrick is typically equipped with an auger that is used to dig the hole for the pole. The proper pole sling is placed around the pole and secured to the winch line. The utility pole is then lifted and positioned into the hole. The pole is then plumbed, and the hole is backfilled and tamped.

Altec, Inc.

Chapter 8 Learner Resources

For additional information, visit
qr.njatcdb.org
Item # 1649

CRANES

Lineworker Rigging Practices

AND DIGGER DERRICKS 9

Altec, Inc.

OBJECTIVES

- Compare the designs, operations, and applications of various types of cranes.
- Identify the general procedures and personnel involved in operating a crane.
- Identify crane dynamics that affect lift safety and describe how to compensate for them.
- Describe the general types of inspections involved in maintaining cranes.

Many different types and sizes of cranes are used for lifting. Some cranes are primarily used for construction work, while others are used for material handling and on-site equipment maintenance, repair, and replacement. Regardless of the type, size, and application, however, many aspects of crane operation, dynamics, and inspections are similar. Also, since cranes and lifting operations involve many potential hazards, safety equipment and procedures are critical.

CRANE TYPES

A *crane* is a combination of a hoist and a structure used to support and move a load. Dozens of different designs are available, with each type adapted to a specific purpose, though only a few are common in construction and industrial settings. Industrial cranes are permanently installed and intended for long-term use. Mobile cranes can be brought to a job site temporarily for a relatively small number of lifts. Tower cranes are semipermanently installed for the duration of a large construction project.

Industrial Cranes

An *industrial crane* is an indoor crane with permanent structural beam supports. Industrial cranes are typically installed in factories, warehouses, or other type of industrial space and are used to move materials and machinery, usually as part of normal operations or maintenance activities. These cranes are composed of a hoist that can travel horizontally on a long beam that is supported by other structural members. Industrial cranes are generally classified based on the type and design of the supporting structural members and include gantry, overhead, and jib cranes.

Gantry Cranes. A *gantry crane* is an industrial crane composed of a hoist trolley that travels along a horizontal bridge beam supported by a leg assembly. **See Figure 9-1.** A *trolley* is a wheeled assembly that travels horizontally along a crane beam or boom. A trolley typically acts as a base for mounting a hoist, providing one direction of horizontal travel for a load.

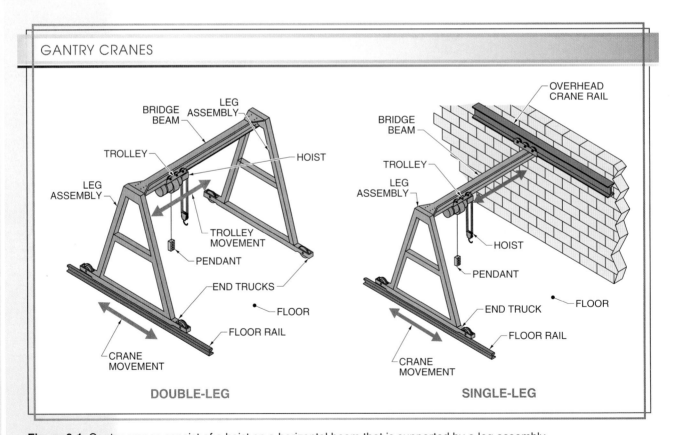

GANTRY CRANES

Figure 9-1. Gantry cranes consist of a hoist on a horizontal beam that is supported by a leg assembly.

Some gantry cranes are affixed to the floor, but many are moved using end trucks. An *end truck* is a roller assembly consisting of a frame, wheels, and bearings. The end trucks of large gantry cranes often travel back and forth on floor rails. Floor rails are small-gauge rails that are recessed into, or set on top of, the floor surface. Between the trolley movement and the crane movement, the hoist can travel in two directions, which allows it to reach any spot within a rectangular area.

A traditional gantry crane design has two sets of legs supporting the ends of the bridge beam. These double-leg gantry cranes run on two parallel floor rails. However, if a gantry crane is to be installed along a straight wall, the design can be modified to use the wall as one of the supports and a single-leg assembly as the other. This is done by fixing a beam to the wall to serve as an overhead crane rail. This type of crane is known as a single-leg gantry crane.

Smaller double-leg gantry cranes that have wheels but do not run on a track are also available. This design allows the gantry crane to be moved easily across any flat floor surface to where it is needed.

Overhead Cranes. An *overhead crane,* also known as a bridge crane, is an industrial crane composed of a hoist trolley that travels along a horizontal bridge beam, which travels along a pair of overhead runways. **See Figure 9-2.** The reach of an overhead crane is anywhere within a rectangle defined by its horizontal lengths of travel. An overhead crane is similar in operation to a gantry crane. The primary difference between the two is the replacement of floor rails with overhead runways.

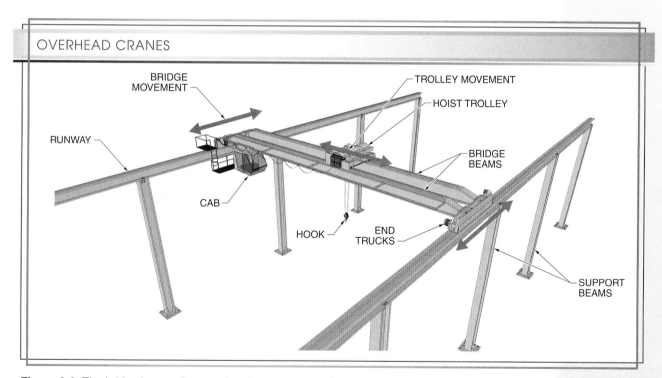

OVERHEAD CRANES

Figure 9-2. The bridge beam of an overhead crane moves along elevated rails called runways. For large overhead cranes, the runways are often built into the structure of a building.

Overhead cranes use different arrangements for trolley travel on the bridge beam and bridge beam travel on the overhead runways. **See Figure 9-3.** A top-running hoist is installed in a trolley that travels along rails on top of a pair of bridge beams. With an underhung hoist, the hoist trolley travels on the upper surface of the lower flange of a single bridge beam.

OVERHEAD CRANE TROLLEYS

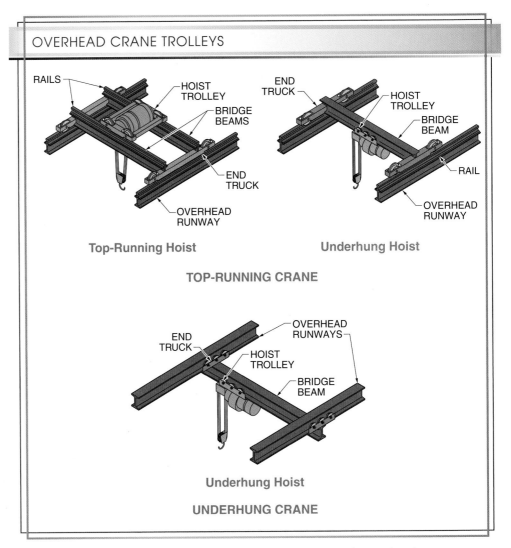

Figure 9-3. Various bridge beam and hoist trolley arrangements for overhead cranes are selected based on their capacity and application.

The bridge-beam assembly can also be either top-running or underhung. End trucks on top-running cranes travel on small-gauge rails mounted on top of the overhead runways. A *runway* is a rail-and-beam combination that allows for the movement of end trucks. A top-running crane with an underhung hoist is the most common overhead crane configuration.

Jib Cranes. A *jib crane* is an industrial crane composed of a hoist trolley that travels along a horizontal boom, which is supported by a single structural leg. A *boom* is a long beam that projects from the main part of a crane in order to extend the reach of the hoist.

The three basic types of jib-crane structures are wall-mounted, base-mounted, and mast. **See Figure 9-4.** Wall-mounted jib cranes are top-braced or cantilevered. A *cantilever* is a projecting structure supported at only one end. Base-mounted jib cranes are freestanding cantilevered cranes mounted on a heavily anchored base. Mast jib cranes have one structural leg, or mast, mounted to both the floor and ceiling and a boom that is cantilevered, underbraced, or top-braced.

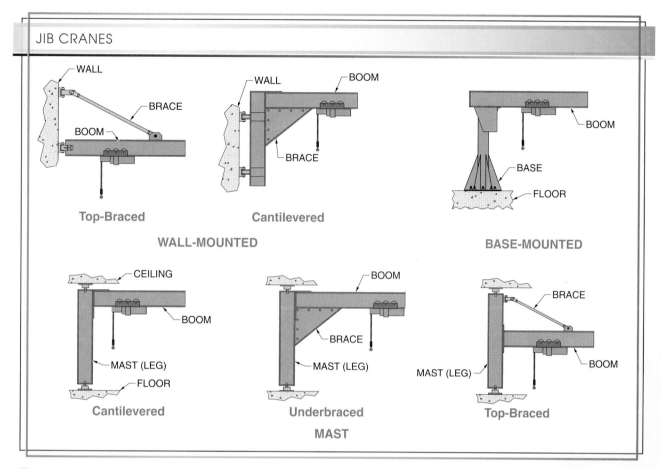

JIB CRANES

WALL-MOUNTED — Top-Braced / Cantilevered

BASE-MOUNTED

MAST — Cantilevered / Underbraced / Top-Braced

Figure 9-4. Jib-crane booms can be configured in several different designs, but all are supported at only one end.

A jib crane may be stationary or capable of rotation. Depending on the mount type, the rotation may be partial, or it may be a full 360°. The trolley travels along the length of the boom. If the jib crane is stationary, the area of reach is limited to the linear boom footprint. If the jib crane is capable of rotation, the area of reach is an arc or full circle.

Due to its cantilevered design, a jib crane is more vulnerable to overloading and damage. Slack should be taken up slowly before a lift to minimize shock to the boom. Rotating the boom on a jib crane should also be done slowly to prevent damage to the load, the surrounding, and individuals.

Mobile Cranes

A *mobile crane* is a crane that can be moved within and between job sites. They are typically used temporarily, brought in for a construction project, and then transported to the next project when finished with the first. Mobile cranes are designed to be small and light enough to be driven over paved roadways, though this design limits their rated capacity.

Mobile cranes are composed of crane assemblies and vehicle platforms. The crane portion includes all hoisting, boom, and operator cab components and is mounted onto the platform on a turntable, allowing the crane to swivel. Many mobile crane variations are possible because the type of platform is relatively independent of the type of crane. **See Figure 9-5.** Therefore, mobile cranes are specified by both the crane type and platform type.

MOBILE CRANE PLATFORMS

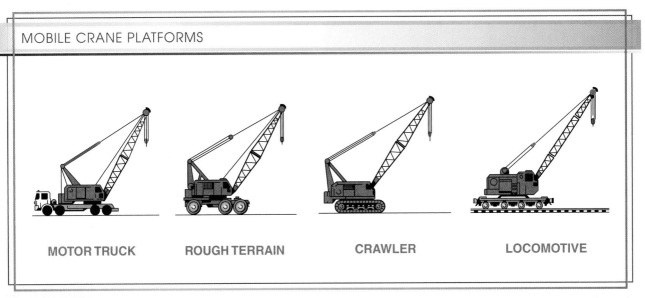

Figure 9-5. Mobile cranes are crane assemblies that are mounted on mobile platforms. The platform used may vary depending on the intended application.

The most common platform type is a motor truck, which is similar to a typical flatbed truck with a long bed. Motor trucks are often used on commercial or residential construction job sites that are accessible by roads.

For off-road use, rough terrain or crawler platforms allow the crane to operate in areas with soft or moderately uneven ground. Rough terrain platforms have large tires with deep treads. Some feature four-wheel steering that allows the crane to move, or "crab," sideways, allowing the crane to operate in tight spaces. Crawler platforms are tracked with linked plates, much like military tanks. These cranes cannot travel on paved roads and must be transported by truck. Cranes are also available mounted on locomotive platforms.

Common types of mobile cranes include telescoping-boom cranes, digger derricks, material-handling bucket trucks, articulating boom cranes, and lattice-boom cranes. These crane types are available on a variety of vehicle platforms.

Telescoping-Boom Cranes. A *telescoping-boom crane* is a mobile crane with an extendable boom composed of nested sections. The crane includes one or two hoists, which are located in the body of the crane behind the boom. Each hoisting rope feeds out over the boom and through a sheave at the far end.

The boom sections are extendable and retractable, allowing for a wide range of boom lengths and changing the distances of the hoist hook from the crane. **See Figure 9-6.** When fully retracted, the crane is easily transported by its vehicle platform, which is a street-legal truck. The boom movement and extension are typically powered by integrated hydraulic systems.

Digger Derricks. A *digger derrick* is a mobile telescoping-boom crane with an auger attachment that is typically mounted on a utility truck. **See Figure 9-7.** Digger derricks are commonly used to dig holes, install utility poles, and lift loads. They can be controlled using fixed controls or remote controls. Digger derricks typically have a lower, intermediate, and upper boom. The digger derrick truck includes four outriggers to stabilize the truck while it lifts a load. Digger derricks may contain other attachments, such as an auxiliary jib, personnel basket, or auxiliary arm.

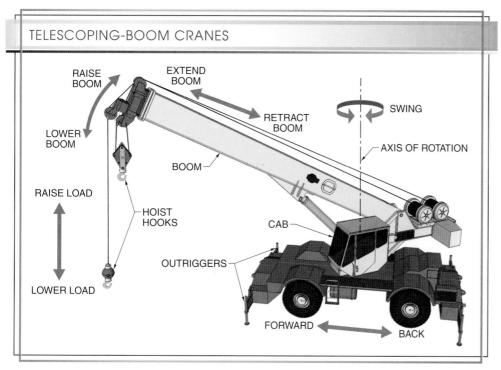

Figure 9-6. A telescoping-boom crane can extend or retract its boom, providing significant versatility to its reach.

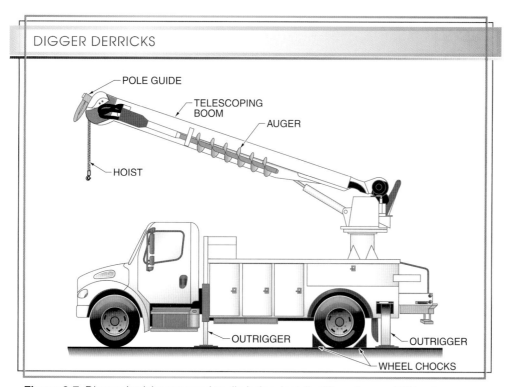

Figure 9-7. Digger derricks are used to dig holes, install utility poles, and lift loads.

Before a digger derrick is used to lift a load, the manufacturer load capacities chart and range diagram must be consulted. **See Figure 9-8.** A *load capacities chart* is the manufacturer-rated maximum weight that a digger derrick can lift under certain circumstances. The load capacities chart is used to determine whether a load can be lifted safely and the combination of booms to extend. A *range diagram* is a manufacturer chart used to determine digger derrick load radius based on the boom angle and extension.

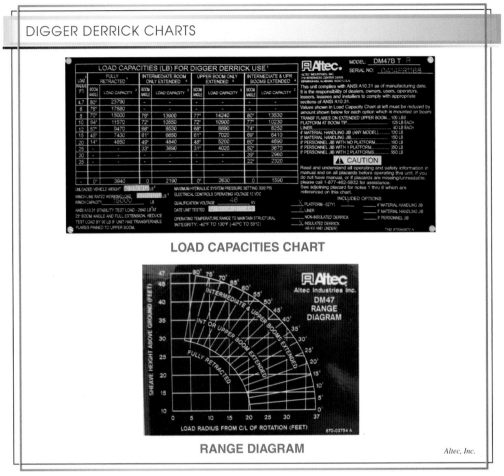

DIGGER DERRICK CHARTS

LOAD CAPACITIES CHART

RANGE DIAGRAM

Altec, Inc.

Figure 9-8. Digger derrick load capacities charts and range diagrams are used to determine whether a load can be lifted safely and which combination of booms to extend.

Material-Handling Bucket Trucks. A *material-handling bucket truck* is an aerial lift with a material-handling jib that is capable of lifting light loads. **See Figure 9-9.** They are typically used to lift personnel to work areas with the added capability of lifting light loads. They may have a telescopic boom or two-part articulating boom.

Articulating Boom Cranes

An *articulating boom crane,* also known as a knuckle boom crane, is a mobile crane that has a boom with a hinged joint to allow the boom to pivot and is typically mounted on commercial trucks. Articulating boom cranes are used to move, load, and unload equipment in a work area or job site. **See Figure 9-10.**

Lattice-Boom Cranes. A *lattice-boom crane* is a mobile crane with a boom constructed from one or more sections of thin steel gridwork. The lattice structure provides a strong boom that is light for its size. **See Figure 9-11.**

MATERIAL-HANDLING BUCKET TRUCKS

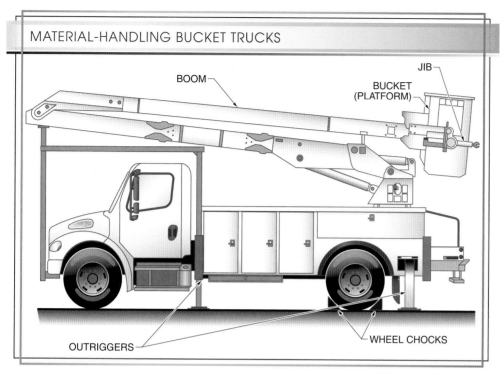

Figure 9-9. Material-handling bucket trucks are aerial (personnel) lifts that contain a material-handling jib that is capable of lifting light loads.

ARTICULATING (KNUCKLE) BOOM CRANES

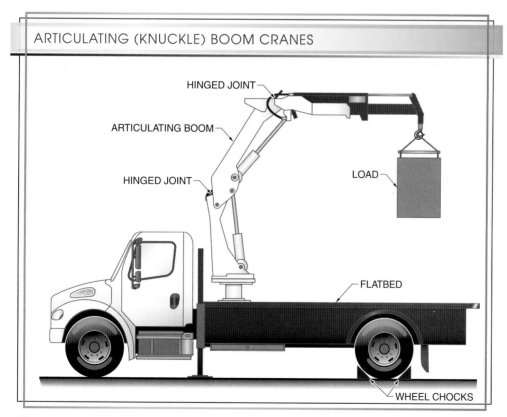

Figure 9-10. Articulating (knuckle) boom cranes are typically used to move, load, and unload equipment in a work area or on a job site.

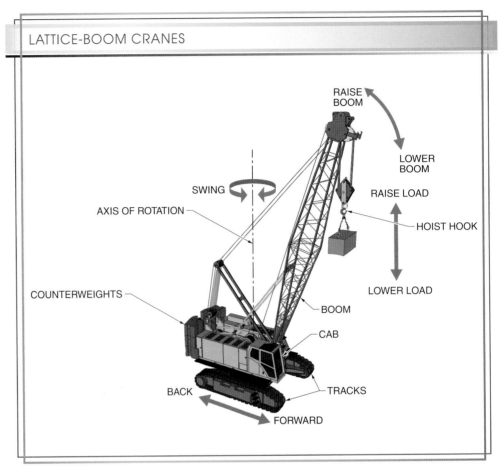

LATTICE-BOOM CRANES

Figure 9-11. The boom of a lattice-boom crane is constructed of lightweight gridwork. Multiple sections can be added in order to provide a variety of boom lengths.

The boom may be composed of single or multiple sections that must be assembled onto the crane body when on the job site. This assembly is complex and requires the use of another, smaller crane. This makes a lattice-boom crane less transportable between job sites, though its vehicle platform allows it to move around while on the site. For this reason, lattice-boom cranes are often deployed for long-term use at large construction sites.

Helicopter Cranes

A *helicopter crane* is a helicopter that is specially designed to lift heavy loads with a long hoist or sling. Helicopter cranes are used to lift loads in remote and inaccessible areas where it is not feasible to use other types of cranes. The outside line industry typically uses helicopter cranes to lift large sections of transmission towers into position and string transmission line. **See Figure 9-12.** The helicopters must comply with all applicable Federal Aviation Administration (FAA) regulations.

Tower Cranes

Tower cranes are commonly used on construction projects such as high-rise buildings. A *tower crane* is a fixed crane consisting of a high, vertical mast tower topped by a horizontal jib. **See Figure 9-13.** The jib has a long arm, which provides horizontal travel for the hoist trolley, and a short arm, which provides the counterweight.

HELICOPTER CRANES

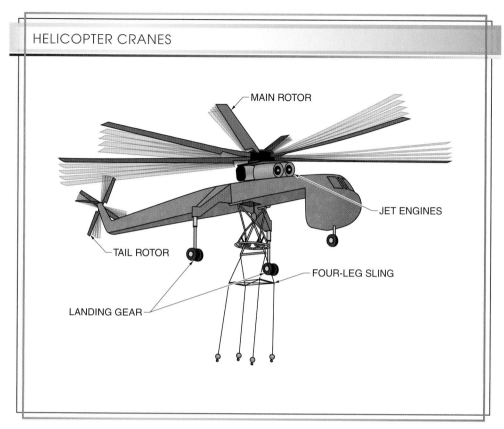

Figure 9-12. Helicopter cranes are used to lift loads in remote and inaccessible areas where it is not feasible to use other types of cranes.

TOWER CRANES

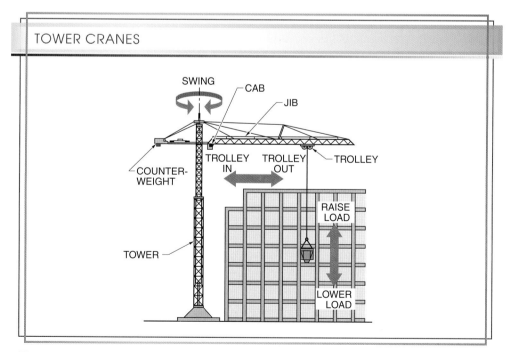

Figure 9-13. Tower cranes are semipermanently installed at a construction site, particularly on high-rise buildings.

The intersection between the mast and jib may be fixed or contain a motorized unit for rotating the jib horizontally. Also, the tower may move along a ground track. This allows the major portion of the crane to be positioned above the job site, saving space.

The operator usually controls the crane from the cabin, which is located at the top of the tower and rotates with the jib assembly. When the crane is not in use, the jib is allowed to rotate freely in order to reduce wind loads on its structure.

Tower crane types are categorized by the way they are attached at the base. A *climbing tower crane* is a tower crane that is secured to the frame of a high-rise structure being erected and can be periodically raised as new floor levels are added to the structure. A *freestanding tower crane* is a tower crane that is secured to a concrete foundation next to the structure being erected. When a tower crane is no longer required on a job site, the crane is dismantled in sections and lowered to the ground. Tower cranes are transported in sections by truck to a job site where they are assembled.

CRANE OPERATION

The operation of a crane to lift and move a load involves many different rigging and lifting considerations, including weight and balance, sling hitch types, rated capacities, safety factors, proper attachment selection, and hoist operation. Cranes use many different types of mechanical advantages to be able to work easily with loads that would otherwise be difficult or impossible to lift or move.

Crane Operators

The crane operator is held directly responsible for the safe operation of the crane. The operator must be properly trained in rigging and lifting procedures and thoroughly familiar with the operation and features of the particular crane model being used. Crane operation requires skill, extreme care, good judgment, alertness, and concentration.

Crane operators must adhere to safety rules and practices as outlined in applicable American National Standards Institute (ANSI) and Occupational Safety and Health Administration (OSHA) standards. In addition, they are typically required to be certified and may be periodically retested for knowledge and skills.

Individuals who cannot speak the appropriate language, read and understand printed instructions, or legally operate construction equipment should not be permitted to operate a crane. Any individual who is hearing or vision impaired or may be suffering from a health condition that might interfere with safe performance should not operate a crane.

Depending on the size and type, cranes are operated from a cab, pendant pushbutton station, or wireless control box. **See Figure 9-14.** A *cab* is a compartment or platform attached to a crane on which an operator rides. A *pendant* is a control unit that hangs down from the hoists of smaller industrial cranes. This pendant includes buttons and/or switches used to control the direction and speed of the various crane movements. A *wireless control box* is a control unit connected wirelessly to a crane or hoist.

Depending on the size and circumstances of the lift, the crane operator may work with other personnel to ensure that safe and proper procedures are followed. Lifting may be performed only after the operator ensures that all rigging, hoisting, and crane components are within specifications. Rigging personnel are in charge of determining the appropriate rigging type, calculating weights and capacities, and selecting the proper equipment. Riggers and signalpersons support the crane operator in placing the lift hook and directing the load travel.

For additional information, visit qr.njatcdb.org Item # 1679

General Lifting Procedures

Lifting procedures vary depending on the particular circumstances, but several general rules apply. A lift should never be attempted if the load is beyond the rated capacities of any of the rigging, lifting, or crane components. Once the hoist is brought directly over the load's center of gravity, the rigger should check that no lines or chains are twisted, overwrapped, or unseated. The rigger then connects the rigging to the hook and ensures that the hook latch is fully closed.

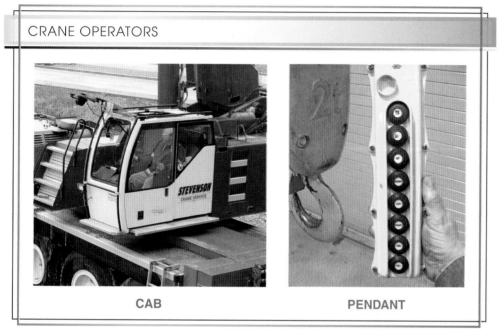

CRANE OPERATORS

CAB

PENDANT

Figure 9-14. Large cranes are typically operated from inside a cab, while small cranes are operated from a pendant that hangs down from the boom or hoist.

The hook is raised slowly until all slack is removed from the rigging. **See Figure 9-15.** The load is then lifted slowly until it is clear of its supports and checked for balance and stability. Crane movements should always be smooth and gradual to avoid shock loading. *Shock loading* is the abrupt application of force to an object. Shock loading and jerky crane movements can cause the load to swing and/or shift, putting extra stress on the rigging or impacting other equipment.

Cranes are often used at construction or industrial sites that involve many people, simultaneous activities, and noisy equipment. It is imperative that all personnel at the job site be alert when working around cranes and other lifting equipment. Personnel not involved in the lift must keep clear of the area around the crane and its load, intended path of travel, and destination.

Signaling

Many cab-operated cranes require the assistance of ground personnel to ensure safe lifting and transportation of the load. In some situations, the crane operator may not have a clear view of the load being hoisted or the final destination of the load. A crane operator then depends upon ground personnel to properly guide the load to its final position. **See Figure 9-16.**

CRANE LIFTS

Altec, Inc.

Figure 9-15. All crane lifts are potentially hazardous situations that require safety equipment and procedures.

SIGNALING

Figure 9-16. Many lifts require the use of a signalperson to direct the crane operator. Commands are often given using standardized hand signals.

Loads should not be moved unless standard crane signals are clearly given, seen, and understood. The operator must pay particular attention to the required crane movements as signaled by the signalperson. The operator takes signals only from the designated signalperson. The only exception to this rule is the Stop signal, which the operator must obey from anyone, at all times.

CRANE DYNAMICS

Crane dynamics are the effects of crane movements and weight on stability and control. The primary concerns are tipping forces called moments. A *moment* is the tendency of a force to rotate an object around a point. With a weight extended out from a crane platform, there is a significant moment (tipping) force on the crane.

Rated Capacities

A crane is typically specified by its rated capacity. The rated capacity is the maximum tension that a crane or hoist may be subjected to while maintaining the appropriate margin of safety. However, this is only the maximum capacity of the crane under certain circumstances. The rated capacity can be temporarily reduced by several factors during most hoisting jobs.

Charts displaying rated capacities that account for all of the applicable factors are available with the crane manufacturer's specifications and on cards located inside the operator's cab. **See Figure 9-17.** Qualified crane operators are trained and tested on how to read these tables. The rigger should work with the crane operator in order to determine whether a load can be lifted safely based on the capacity charts and lift conditions.

RATED CAPACITY CHARTS

Boom Length (ft)	Radius (ft)	Boom Angle (°)	Maximum Allowable Loads (lb)		
			With Outriggers	Without Outriggers	
				Over Side	Over Rear
40	12	73	220,000	100,400	139,700
	15	68	185,000	71,800	97,400
	20	60	149,000	48,100	54,600
	25	51	118,000	35,800	47,900
	30	41	86,000	28,300	37,800
	35	29	67,000	23,200	31,100
60	15	76	180,000	71,000	97,000
	20	71	147,000	47,300	63,900
	25	65	118,500	35,000	47,200
	30	60	86,200	27,500	37,100
	35	54	67,400	22,400	30,300
			54,700		5,500

Figure 9-17. Crane rated capacities are provided in charts that list relevant lift factors.

Many cranes include an alarm that will sound if the attempted lift weight exceeds the rated capacity for the current boom length and orientation. This alarm must never be ignored. Overloading a crane can damage the boom or cause the entire crane to tip over, which can result in injuries and property damage.

Load Radius. As a load gets farther from the crane's base, the moment force on the crane increases. At some distance, the load will cause the crane to tip over. **See Figure 9-18.** Therefore, the rated capacity of a crane depends on the maximum load radius during a lift. The *load radius* is the horizontal distance from the pivot point of a crane to a point below the hoist hook.

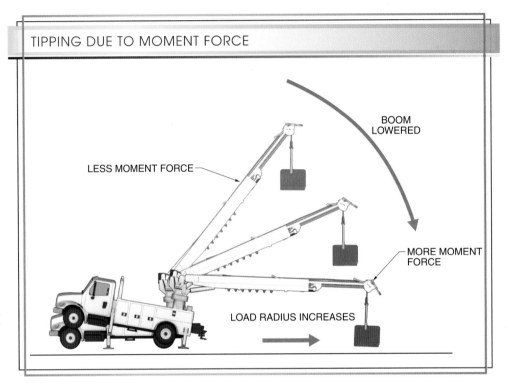

TIPPING DUE TO MOMENT FORCE

LESS MOMENT FORCE

BOOM LOWERED

MORE MOMENT FORCE

LOAD RADIUS INCREASES

Figure 9-18. As a load moves farther out from the crane, the moment (tipping) force increases and can eventually cause the crane to tip over.

For example, a certain load located 20′ from the crane must be moved to a new location 40′ away. Since the farther distance reduces the rated capacity of the crane, it determines the maximum allowable load weight for the entire lift. The crane's rated capacity may be 90 t at 20′ but only 60 t at 40′. Therefore, the maximum load for this particular lift is only 60 t.

It should be noted, however, that the greatest radius may be at any point in the lift. If the load must be extended out to a radius of 50′ during the lift, such as to avoid an obstacle, before returning to a radius of 40′, then that farther distance should be used to determine the maximum load allowed.

The load radius is determined by the boom angle and boom length. *Boom angle* is the angle between a horizontal plane and a boom. Greater boom angles result in a reduced working radius and a higher rated capacity (if the boom length remains the same). As a boom is lowered toward the horizontal, the load radius increases, but the rated capacity decreases. Boom length also changes the load radius. **See Figure 9-19.**

LOAD RADIUS

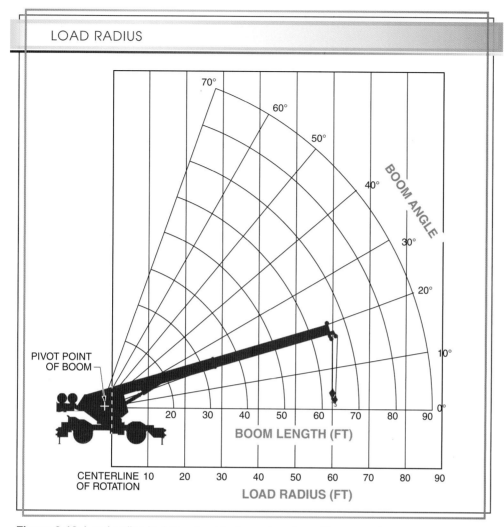

Figure 9-19. Load radius is determined by boom length and boom angle.

Boom length is constant for some cranes but can change for others. For example, telescoping-boom cranes can extend or retract their booms during a lift. Also, the hoist hook may not be mounted on the end of a boom. Some cranes have secondary hoists that drop the hook from sheaves located along the length of the boom. Also, a tower crane trolley can be at any point along the jib (boom). In such cases, the load radius is measured only to the hook, not to the end of the boom.

Counterweights. A *counterweight* is a block of dense material attached to one end of a structure in order to offset the moment force of the weight attached to the other end of the structure. Crane counterweights are typically modular, platelike pieces of steel or concrete mounted opposite the boom. **See Figure 9-20.**

Counterweights can be added to, or removed from, the crane to change the crane's rated capacity. Additional counterweights increase the rated capacity. However, this is effective only to a point, as increasing the number of counterweights beyond the recommended limit puts excessive stress on the crane platform and increases the weight of the crane, making it more likely to sink into the ground.

COUNTERWEIGHTS

COUNTERWEIGHTS

Figure 9-20. Counterweights are heavy blocks of metal or concrete attached to the back of a crane body to counteract the moment force during a lift.

Outriggers. An *outrigger* is a structure that extends out from the platform of a mobile crane and contacts the ground with a large pad. **See Figure 9-21.** An outrigger increases the effective footprint of a crane, which makes tipping less likely. It also reduces the pressure a crane exerts on the ground by increasing its surface contact.

Outriggers are hydraulically operated and thus they can raise the body of a crane so that it is supported only by the outrigger pads. This ensures full, even contact with the ground. Outrigger effectiveness can be improved, particularly on soft ground, by placing larger plates or timbers under the pads.

Mobile cranes may use an "on rubber" set of rated capacities for hoisting and traveling with loads when the crane is on its tires only. When performing these types of operations, the boom typically must be centered over the front of the crane within a narrow angular range. The rated capacity for this configuration is significantly lower than for most other operations.

Ground Stability. When positioning a crane or its outriggers, the stability and integrity of the ground under the crane is critical. A crane should never be positioned over voids, chambers, or piping that might collapse under its weight. Unstable soil caused by improper compaction, water infiltration, or excessive slope can give way, causing the crane to collapse. A thorough site survey should be conducted prior to deciding upon the placement of a crane in order to avoid these hazards.

Boom Direction. The direction of the boom in relation to the platform affects stability because cranes are typically not as wide as they are long. The rated capacity of a crane is typically less when the boom is over the side of the crane than when it is over the rear or front. These areas in which a crane can operate are known as "quadrants of operation." **See Figure 9-22.** Outriggers compensate for this difference to some degree because they typically extend from the sides of the crane, making the footprint wider. However, a crane's capacity may still be somewhat reduced.

OUTRIGGERS

Altec, Inc.

Figure 9-21. Outriggers increase the footprint of a crane, making it more stable and increasing its rated capacity.

QUADRANTS OF OPERATION

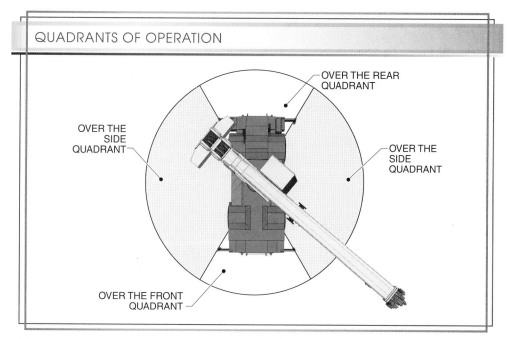

OVER THE REAR QUADRANT

OVER THE SIDE QUADRANT

OVER THE SIDE QUADRANT

OVER THE FRONT QUADRANT

Figure 9-22. Lift capacity may vary in different quadrants of operation because the crane is more stable in its long direction (over the front and over the rear) than in its short direction (over the sides).

Moving Loads Horizontally

During a lift, it is common to move a load horizontally. This is done easily with industrial and tower cranes, but the angled-boom designs of mobile cranes make this movement more complicated. Changing the boom angle moves the load not only horizontally, but also vertically, which may not be desirable.

For cranes with fixed-length booms, such as a lattice-boom crane, the vertical movement can be countered by simultaneously raising or lowering the hoist. **See Figure 9-23.** There is even a pair of special crane signals that specify this combination. However, the range of horizontal movement is still limited with this type of crane. Plus, the boom-angle change required is significant and may present clearance issues.

MOVING LOADS HORIZONTALLY

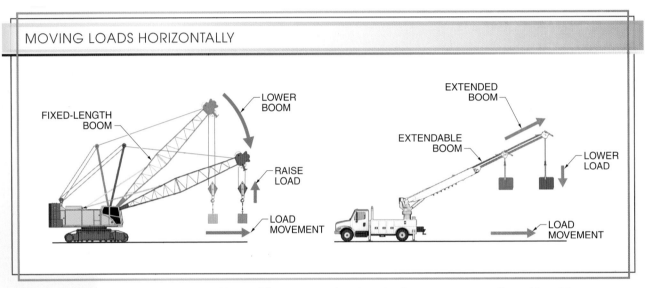

Figure 9-23. Moving loads horizontally with a mobile crane requires two simultaneous crane motions. All mobile cranes can move loads horizontally, though telescoping-boom cranes typically have a greater horizontal range.

A telescoping-boom crane, however, is much more versatile in this regard. The ability to change the boom length provides a greater range of horizontal movement. A combination of extending the boom while lowering the load moves the load outward horizontally, and retracting the boom while raising the load moves a load inward. These movements are particularly useful for placing a load between the floors of a building under construction.

As always, the change in load radius from these movements affects the crane's rated capacity for the lift. A careful analysis should be done prior to the lift to ensure that the rated capacity will not be exceeded in any of the positions to be used.

Boom Deflection

Boom deflection is the bending down of a boom due to the weight of a load. Booms are designed to be somewhat flexible, so some amount of bending is expected. As long as the load weight is less than the crane's rated capacity, then the bending should be within normal specifications. However, this bending may still need to be compensated for during hoisting.

Boom deflection may occur at the beginning of a lift as slack is taken up from the rigging and the weight of the load starts to be supported by the crane. This bending moves the end of the boom outward and downward slightly, causing the boom and the load to become misaligned. **See Figure 9-24.** If the load is hoisted from the ground without first correcting this misalignment, the load will swing outward, away from the crane. When the deflection is significant, the resulting swinging may constitute a loss of control of the load and present a hazard for ground personnel.

BOOM DEFLECTION

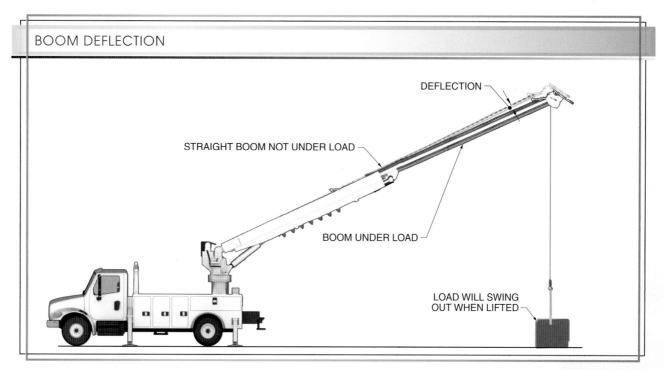

DEFLECTION

STRAIGHT BOOM NOT UNDER LOAD

BOOM UNDER LOAD

LOAD WILL SWING OUT WHEN LIFTED

Figure 9-24. Minor deflection of a boom as it comes under load at the beginning of a lift is normal but may result in the load being lifted at an angle. This can cause the load to swing, which can cause damage or present a safety hazard.

To correct this problem, the signalperson must first observe whether boom deflection is occurring, and then signal the crane operator to apply the appropriate corrective action. Just as the load is about to be lifted off the ground, the hoisting line should be checked to see if it is vertical. If it is not, the signalperson should give the signal to raise the boom and lower (hold) the load or, if the boom cannot be raised, to retract the boom until the hoisting line is vertical. Once the hoisting line is vertical, the load should not swing when hoisted.

Side Loading

Cranes are designed to lift loads in a vertical direction only. If a crane is not level or a load is pulled from the side, the load weight can cause side loading on the boom, which can excessively stress and damage it. *Side loading* is the application of a horizontal force on a boom due to the hoisting of a load that is not directly underneath the hoist hook.

A quick visual check can determine whether the crane is level. The free-hanging winch line is viewed from head on to see if it is aligned with the centerline of the boom. **See Figure 9-25.** If the cable does not line up with the boom, the crane should be releveled until the proper alignment is achieved. This check should be performed twice, with the boom in two positions, 90° apart. This ensures that the crane is level in all directions.

Side loading of the boom may also occur if an attempt is made to drag a load, or if the boom sheave is not directly over the load. Dragging a load with a crane is dangerous and may cause damage or an immediate failure of the boom. If the hoist hook cannot be positioned directly over the load lift point, either the crane or the load must be relocated.

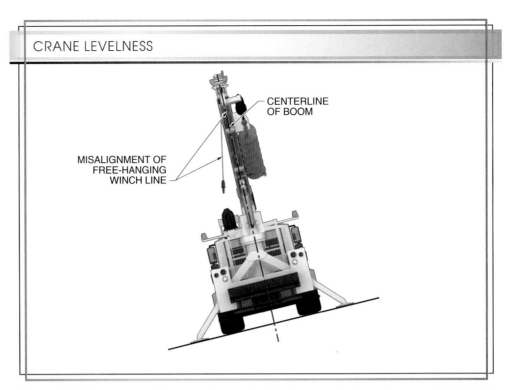

CRANE LEVELNESS

CENTERLINE
OF BOOM

MISALIGNMENT OF
FREE-HANGING
WINCH LINE

Figure 9-25. Cranes should be visually checked for levelness by the alignment of the boom and a free-hanging winch line.

Drift

If a load is moved too rapidly, it will drift. *Drift* is the tendency of a load to continue moving after the crane has stopped. **See Figure 9-26.** Drift can be caused in mobile cranes by swinging or extending the boom rapidly and in industrial and tower cranes by moving the trolley rapidly. Drift causes side loading of the boom/trolley and a potentially hazardous loss of control of the load.

The crane operator can compensate for drift at the end of a crane movement with a small, quick movement in the opposite direction. However, when not done properly, this can make the problem worse. Instead, the best solution to prevent drift is to move the load slowly and use tag lines to help control the load.

Wind Loads

Wind loads can be a significant factor in load stability and safe crane operation. Crane manufacturers often have recommended safe wind speeds, above which all hoisting operations must cease. The maximum wind speed varies for different types of cranes.

If there is no wind speed recommendation by the manufacturer, a competent person should evaluate the lift conditions for safety. Wind-speed-indicating devices are mandated by OSHA under certain circumstances to assist the crane operator in determining whether the wind speeds are too high to make a lift.

Loads with large surface areas relative to their weight are more susceptible to wind loads, and extra caution should be used when hoisting such loads. Wind can also affect the crane directly by putting pressure on the boom. Depending on the wind direction, wind can add to the moment force and potentially cause the crane to lose stability. The lift should be postponed if there are indications that wind loads will become a problem, even if the wind speed is normally safe.

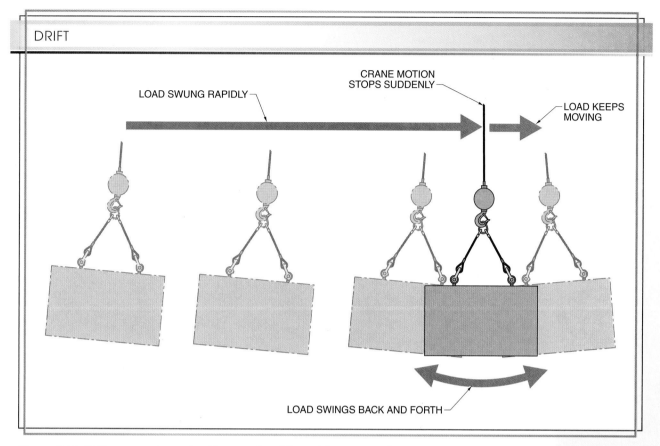

DRIFT

LOAD SWUNG RAPIDLY

CRANE MOTION
STOPS SUDDENLY

LOAD KEEPS
MOVING

LOAD SWINGS BACK AND FORTH

Figure 9-26. Drift occurs when a load is moved rapidly and then stopped.

Two Blocking

Two blocking is a condition in which a device on the hoisting line, such as a block or overhaul ball, is drawn up against the sheave at the end of a boom. If the hoisting line is drawn up further, it may be broken because of the mechanical interference of the parts at the end of the boom. An *anti-two-block device* is a safety device mounted close to the fixed end of a crane's hoisting line that either sounds an alarm or immediately stops the crane and hoist if it is touched by the hoist hook assembly. **See Figure 9-27.** This prevents damage from drawing the boom end components too close together.

Although most cranes are equipped with the device, an anti-two-block device should not be relied upon to prevent breakage of the hoisting line. The signalperson should pay attention to the position of the block relative to the end sheave and signal the crane operator to lower the hoist to prevent contact between the components.

Power Line Clearances

Any time a crane is to be erected, dismantled, or operated near an energized power line, special precautions must be taken. OSHA prescribes a number of options for preventing contact with energized power lines, some of which must be used in conjunction with other options.

The first option is to have the utility company de-energize and ground the power lines. All power lines must be presumed energized until the utility owner/operator confirms that they have been de-energized and grounded. When this option is used, the grounding means must be present and visible at the job site so that its integrity and presence can be observed. After de-energizing and grounding are accomplished, lifting operations can proceed normally. However, the crane operator, signalperson, and rigging crew must still make sure that the crane or load does not snag or come in contact with the power lines.

The second option is to ensure that no part of the crane, load line, or load comes within a certain distance of the energized power line. The minimum clearance distance for safe operation varies according to the voltage of the power line. The second option specifies a set distance of 20′ for voltages up to 350 kV, or 50′ for more than 350 kV up to 1000 kV.

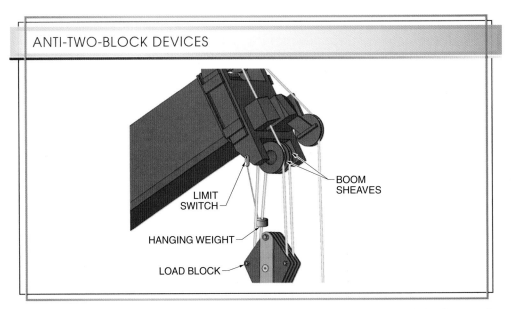

ANTI-TWO-BLOCK DEVICES

LIMIT SWITCH

BOOM SHEAVES

HANGING WEIGHT

LOAD BLOCK

Figure 9-27. An anti-two-block device is a safety device designed to prevent hoist damage if the hoist hook assembly gets close to the sheave or block at the end of a boom.

If the power line voltage is known more precisely, however, a third option may allow for the smaller clearance distances as given in Table A of OSHA Subpart CC, 1926.1408. **See Figure 9-28.** For voltages over 1000 kV, the minimum clearance distance must be determined by a person qualified with respect to power transmission and distribution.

If either the second or third option is used, a planning meeting must be held with the operator and others who will be working within the vicinity of the hoisting equipment and the load. A review of the location of the power lines and steps that will be implemented to prevent encroachment and electrocution must be discussed as part of the meeting.

Elevated warning lines, barricades, or a line of signs must be used to warn the crane operator of the safe working boundary. **See Figure 9-29.** If the operator cannot see these warning devices, a dedicated spotter who is in continuous contact with the operator must be used to assist in maintaining the required clearance distance. Painted lines on the ground, elevated stanchions, or line-of-sight landmarks may be used to help the spotter. In addition to the spotter, a secondary means of protection, such as a proximity alarm, range control warning device, range limiter, or insulating link, must also be employed. Also, any tag lines used to control the load must be nonconductive.

CRANE INSPECTIONS

A competent person must visually inspect a crane for proper operation and safety controls. This must be done monthly and also prior to each shift during which the equipment will be used. This inspection involves looking for apparent problems. Taking equipment apart is not required as part of this inspection unless the visual inspection or trial operation indicates that further investigation is needed. In this case, an immediate determination must be made by the competent person as to whether the deficiency constitutes a safety hazard. If so, the crane must be taken out of service until the deficiency has been corrected.

MINIMUM CLEARANCE DISTANCES TO ENERGIZED POWER LINES

Voltage*	Minimum Clearance Distance†
Up to 50	10
50 to 200	15
200 to 350	20
350 to 500	25
500 to 750	35
750 to 1000	45

* in kV AC, nominal
† in ft

Figure 9-28. When assembling, operating, or disassembling a crane near energized power lines, minimum clearances must be maintained for safety. The required distance depends on the power line voltage.

WORKING NEAR POWER LINES

Figure 9-29. Perimeter-alerting safety measures are used to help crane operators maintain the required safe distance from energized power lines.

Crane inspections fall roughly into three areas: the platform, the hoist positioning components, and the hoist. Each area can include complex machinery that requires many different types of inspections. The inspection requirements in each area vary depending on the type of crane and its application, though some basic guidelines are generally applicable.

Platform Inspections

The platform includes the structural support of the crane and any related floor or wall components. On a mobile crane, the platform also provides mobility. The platform should be inspected for corrosion, distorted members, missing fasteners, and damaged welds. Operator cab windows should be checked for significant cracks, breaks, and other deficiencies that would hinder the operator's view. Rails along the floor or walls should be checked for clean, smooth surfaces and any damage from other equipment. Mobile platforms are subject to any applicable vehicle maintenance and inspection activities, such as engine maintenance and tire inspection.

Ground conditions around the crane should be inspected for proper support, including ground settling under and around outriggers and supporting foundations, ground water accumulation, or similar conditions. The crane should be checked for level position within the tolerances specified by the equipment manufacturer's recommendations, both before each shift and after each move and setup.

Positioning Component Inspections

The positioning components include booms, bridge beams, hoist trolleys, and end trucks. Wheels should be checked for flat spots, cracks, and breaks. The wheel bearings should be checked for proper lubrication, intact seals, and physical damage.

Trolley inspections should include checking the dimension between wheel flanges. **See Figure 9-30.** On most hoist trolleys, the dimension between the flange landings of both wheels must be between ⅛″ and ¼″ greater than the beam flange width. The addition or removal of spacing washers may be necessary to obtain the proper dimension while keeping the hoist centered under the bridge beam. Some hoist trolleys are not adjustable.

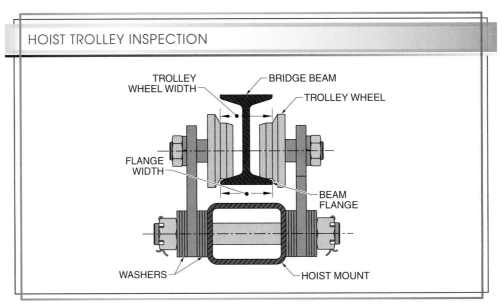

Figure 9-30. Trolley inspection involves checking the trolley for the necessary gap between the wheels and the bridge beam.

Boom systems include complex actuation and control mechanisms, such as hydraulic cylinders. Control mechanisms should be checked for maladjustments interfering with proper operation. Control and drive mechanisms must not have apparent excessive wear or contamination by lubricants, water, or other foreign matter. Inspections should include checking for low fluid levels, dirty filters, leaks, and stiff or unresponsive controls. More intensive periodic inspections may require the expertise of personnel who specialize in these power systems.

Hoist Inspections

Though they are mounted in different ways, the basic components of a hoist are similar in all applications. The manufacturer's literature should be referred to for any cleaning or lubrication requirements. Any unusual noises, jerky operations, or unresponsive controls indicate problems with the hoist.

Wire rope should be inspected for wear and damage. Proper wire rope reeving and drum wrap should be checked, and hoist brakes should be tested for quick response. The hook and its latch must be inspected for deformation, cracks, excessive wear, and damage from chemicals or heat.

Crane operators, assistants, and other personnel involved in a lift should be alert for any problems with any of the equipment, both before and during the lift. If any problems are observed or suspected, the load and crane controls should be placed in safe positions and a supervisor should be immediately notified.

COMMON APPLICATIONS— INSTALLING SCREW ANCHORS

A power-installed screw anchor (PISA) is a screw anchor that is installed in the ground with the assistance of a digger derrick. PISAs are typically used for anchoring guy wires. The anchor must be installed to a sufficient depth to provide a solid hold. The anchor must be set into the ground on an angle in line with the pull of the guy wire.

To install an anchor with a digger derrick, apply the following procedure:
1. Set up the digger derrick so that the boom is in line with the screw anchor.
2. Remove the auger from the digger derrick and install the anchor wrench assembly.
3. Attach the anchor rod to the wrench assembly.
4. Install the screw anchor into the ground.
5. Remove the anchor wrench and install the eye-nut on the anchor rod.

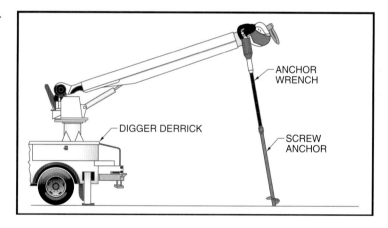

For additional information, visit
qr.njatcdb.org
Item # 1649

Lineworker Rigging Practices

LIFT PLANNING 10

- Differentiate between common and complex lifts.
- Describe the development and purpose of a lift plan.
- Explain the considerations in evaluating a job site and designated crane area for safety.
- List and describe the factors that affect a planned load path.
- Describe the common considerations of load preparation.
- Describe the two aspects of load control and how they are addressed.
- Compare the usage and characteristics of various material-handling equipment.
- Describe the general procedure for performing a prelift check.

Southern California Edison

All rigging and hoisting tasks require evaluation and planning prior to a lift. This includes relatively simple lifts even though the lift plan and equipment involved may be minimal. More complex lifts may involve multiple cranes or hoists, specialized lifting devices, and specially trained personnel. These more complex lifts require written lift plans that fully evaluate all conditions and potential hazards. The handling of loads immediately before and after a lift must also be considered.

COMMON AND COMPLEX LIFTS

Lifts are generally regarded as either common or complex. This exact terminology may vary, but the outside line industry generally recognizes a difference in the preparation required for relatively simple lifts and for more critical or more complicated lifts. However, the definitions of these two categories are not standardized.

As a general rule, common lifts involve loads with well-defined lift points that can be safely hoisted with one or two slings and either a single crane or one or two manual hoists. No written plan or special precautions are usually required for common lifts.

Any lift other than a common lift is typically considered a complex lift. A lift may also be considered complex due to a number of factors. The factors may be explicitly defined by the facility or contractor, or they may be determined by the personnel involved based on their experience. However, the factors are always related to aspects of the lift that increase hazards, require multiple steps, require additional personnel, involve extremely heavy loads, take place in an inherently hazardous location, and/or require workers to operate under unusual constraints.

If there is any doubt whether a lift should be considered common or complex, it is best practice to proceed as if it is complex. Examples of factors that affect the complexity of a lift include the following:

- requiring multiple movements of the load
- hoisting extremely heavy loads
- affecting nearby equipment or processes
- requiring personnel to be lifted
- hoisting a load with more than one crane or powered hoist simultaneously
- turning or shifting a load while hoisted
- rigging for sling angles of less than 30° from horizontal
- rigging with attachment points that are below the load's center of gravity
- hoisting a load where the center of gravity is unknown
- hoisting within confined spaces
- hoisting in a nuclear facility
- requiring special rigging skills

The rigging, signaling, and equipment-operating personnel used for complex lifts must have more extensive experience and higher levels of training than those involved with common lifts. These requirements should be included in a lift plan.

LIFT PLANS

Complex lifts and some common lifts require the development of a lift plan. A *lift plan* is an evaluation of the potential hazards of a lift and the equipment and procedures required to safely execute the lift. A lift plan is used by engineering, safety, field, and any other necessary personnel to prepare for a lift.

The development of a lift plan involves identifying potential problems and then developing appropriate solutions for those problems. A particular problem may have a number of different solutions, many of which could be acceptable. Many personnel may be involved in evaluating potential solutions and choosing the best one.

Lift plans for some lifts may be discussed relatively informally. For example, a brief meeting between the rigger, the signalperson, and the crane operator may be adequate to review a relatively simple lift procedure and address all safety concerns.

However, a complex lift requires more extensive planning and evaluation. A complex lift plan is likely to require documentation of the load characteristics, the load path, and load control procedures. However, there is no single document format applicable to all types of lifts. Rather, contractors or facilities may have a template to use as a starting point and then attach additional specification sheets, drawings, and other information as appropriate. **See Figure 10-1.** If the lifting activity will be performed on a periodic basis, lift plan documentation should be saved and updated as needed for future use.

EXAMPLE LIFT PLAN FORM

Lift Plan
(Over 5,000 lbs)

LOCATION:_____ DATE OF LIFT:_____
LOAD DESCRIPTION_____
LIFT DESCRIPTION _____

A. WEIGHT
1. EQUIPMENT CONDITION   NEW ☐   USED ☐
2. WEIGHT EMPTY _____ LBS
3. WEIGHT OF HEADACHE BALL _____ LBS
4. WEIGHT OF BLOCK _____ LBS
5. WEIGHT OF LIFTING BAR _____ LBS
6. WEIGHT OF SLINGS & SHACKLES_____ LBS
7. WEIGHT OF JIB
   ☐ ERECT   ☐ STORED
8. WEIGHT OF HEADACHE BALL
   ON JIB _____ LBS
9. WEIGHT OF CABLE (LOAD FALL) _____ LBS
10. ALLOWANCE FOR UNACCOUNTED
    MATERIAL IN EQUIPMENT _____ LBS
11. OTHER _____ LBS
           TOTAL WEIGHT _____ LBS

SOURCE OF LOAD WEIGHT

_____
(Name Plate, Drawings., Calculated, etc.)
WEIGHTS VERIFIED BY:

_____

B. JIB
ERECTED _____ STORED

1. IF JIB TO BE USED _____
2. LENGTH OF JIB _____
3. ANGLE OF JIB _____
4. RATED CAPACITY OF JIB (FROM CHART) [____]

C. CRANE PLACEMENT
1. ANY DEVIATION FROM SMOOTH SOLID FOUNDATION
   IN THE AREA?

_____

2. ELECTRICAL HAZARDS IN AREA?

_____

3. OBSTACLES OR OBSTRUCTIONS TO LIFT OR SWING?

_____

4. SWING DIRECTION AND DEGREES (BOOM SWING)

_____

D. CABLE
1. NUMBER OF PARTS OF CABLE _____
2. SIZE OF CABLE _____

E. SIZING OF SLINGS
1. SLING SELECTION
   A. TYPE OF ARRANGEMENT _____
   B. NUMBER OF SLINGS IN HOOKUP _____
   C. SLING SIZE _____
   D. SLING LENGTH _____
   E. RATED CAPACITY OF SLING: [____]

2. SHACKLE SELECTION
   A. PIN DIAMETER (INCHES) _____
   B. CAPACITY (TONS) _____
   C. SHACKLE ATTACHED TO LOAD BY: _____
   _____
   D. NUMBER OF SHACKLES _____

F. CRANE
1. TYPE OF CRANE _____
2. CRANE CAPACITY _____ TONS
3. LIFTING AGREEMENT
   A. MAXIMUM DISTANCE CENTER OF LOAD TO CENTER
      PIN OF CRANE _____
   B. LENGTH OF BOOM _____

   C. ANGLE OF BOOM AT PICKUP _____ DEGREES
   D. ANGLE OF BOOM AT SET _____ DEGREES
   E. RATED CAPACITY OF CRANE UNDER SEVEREST
      LIFTING CONDITIONS (FROM CHART)
      1. OVER REAR _____ LBS
      2. OVER FRONT _____ LBS
      3. OVER SIDE _____ LBS
      4. FROM CHART - RATED CAPACITY OF
         CRANE FOR THIS LIFT [____]

5. MAXIMUM LOAD ON CRANE _____
6. LIFT IS [____] % OF CRANE'S
   RATED CAPACITY

G. PRE-LIFT CHECKLIST   YES   NO
1. MATTING ACCEPTABLE   ☐   ☐
2. OUTRIGGERS FULL EXTENDED ☐   ☐
3. CRANE IN GOOD CONDITION ☐   ☐
4. SWING ROOM   ☐   ☐
5. HEAD ROOM CHECKED ☐   ☐
6. MAX COUNTERWEIGHTS USED ☐   ☐
7. TAG LINE USED ☐   ☐
8. EXPERIENCED OPERATOR ☐   ☐
9. EXPERIENCED FLAGMAN ☐   ☐
   (DESIGNATED)
10. EXPERIENCED RIGGER ☐   ☐
11. LOAD CHART IN CRANE ☐   ☐
12. WIND CONDITIONS _____
13. CRANE INSPECTED BY _____
14. FUNCTIONAL TEST OF CRANE BY _____

SPECIAL INSTRUCTIONS OR RESTRICTIONS FOR CRANE, RIGGING, LIFT, ETC.

MULTIPLE CRANE LIFTS REQUIRE A SEPARATE LIFT PLAN FOR EACH CRANE. ANY CHANGES IN THE CONFIGURATION OF THE CRANE PLACEMENT, RIGGING, LIFTING SCHEME, ETC. OR CHANGES IN ANY CALCULATIONS REQUIRE THAT A NEW LIFT PLAN BE DEVELOPED.

_____ DATE _____   _____ DATE _____
Signature of Task Supervisor   Signature Plan Checked by Rigging Supervisor

Figure 10-1. A lift plan form provides a checklist of common considerations during a lift. Additional documentation and drawings may also be needed, but the form provides a starting point for planning.

Lift planners should be qualified persons who are familiar with the relevant federal, state, provincial, and industry safety standards and regulations that pertain to the rigging, hoisting, and material-handling tasks to be performed. The services of a qualified engineer are necessary if the lift requires equipment or procedures that are unusual and do not conform to established standards.

Site Evaluation

A job site where cranes will be used should be evaluated prior to the equipment arriving at the site. Many cranes require a firm, graded, and well-drained area for initial assembly and operation. Some job sites may even require the building of temporary roads and staging areas. **See Figure 10-2.**

TEMPORARY ROADS

Figure 10-2. A job site may require temporary roads and staging areas. This information may be part of a lift plan.

Any underground voids, tanks, and utilities should be identified and considered along with any overhead power lines, elevated process pipelines, and other above-ground obstructions so that appropriate clearance distances can be maintained. The required clearances may be based on government, industry, and facility requirements.

Ground conditions should be evaluated to determine whether the ground can support the weight of the hoisting equipment and load. The services of a qualified civil engineer may be required to determine whether there is sufficient load-bearing capacity or if additional blocking, matting, or other support is needed. Locations with significant slope, noncompacted or unstable soils, or underground voids may require a particular type of crane. Also, potential changes in site conditions should be considered. For example, a site that may be suitable in dry conditions may be unsafe if the ground becomes wet. The appropriate preventive measures should be based on a worst-case scenario.

Staging areas may have to be cleared for the delivery of crane components. It may also be necessary to clear staging areas for the assembly of lattice boom or tower cranes. Grading, shoring, or blocking may be required in order to keep components from shifting during storage.

Crane Clearances

The site must have sufficient space for the setup and safe operation of the crane. If the crane needs to be assembled on site, space must be available for the crane sections and the assembly crane, if needed. There must also be enough space for the maximum footprint of the crane and its surrounding safe zone while in operation. Extended outriggers and any additional space needed for blocking under the outrigger pads must be considered. **See Figure 10-3.**

CRANE CLEARANCES

Figure 10-3. Various clearances are required around an operating crane in order to keep personnel and equipment safe.

The evaluation of working space around the crane is particularly important because there is typically less space available once a job site becomes busier and materials are stored in the area. Space around the crane's upperworks should be evaluated to determine whether there is enough clearance to avoid pinch points or interference as the crane swings.

Load Paths

Many loads can be lifted and moved in a single operation without undue risk to personnel or equipment. These common lifts can usually be performed without an extensive evaluation of the load path. A *load path* is the path a load must take from its starting point to its destination and that includes every point in between. **See Figure 10-4.** Complex lifts, however, require site plans, measurements, and elevation drawings that are then used to develop sketches or drawings of the load path.

Fall Zones. A *fall zone* is an area where partially or completely suspended materials might land if they become loose from a load or if the rigging or crane fails. All fall zones should be identified as part of the lift-planning process. The size of a fall zone depends

LOAD PATHS

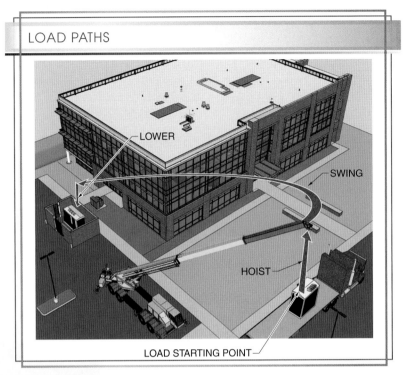

Figure 10-4. A load path includes the load's starting point, destination, and every point in between. The load path helps plan for any clearances, obstructions, and prohibited areas.

on the planned height of the load when it is overhead. **See Figure 10-5.** Ideally, a load should be moved while keeping it as close to the ground as possible, which minimizes the fall zone area needed. However, as the load is hoisted, the area threatened by a load fall becomes larger.

Warning signs, barricades, or caution lines may be necessary in order to prevent personnel from entering a fall zone. Certain areas of a job site may be temporarily evacuated as loads are hoisted overhead in order to keep workers safe. As a rule, only the personnel needed to connect or receive a load are permitted to be within the fall zone as a load is hoisted or lowered into place.

Obstructions. Obstructions may interfere with the direct path of a load from one point to another. It may be necessary to change load direction, height, or orientation during a lift in order to avoid obstacles. In complex lifts, it may even be necessary to set the load down temporarily and reposition the crane in order to avoid obstacles.

FALL ZONES

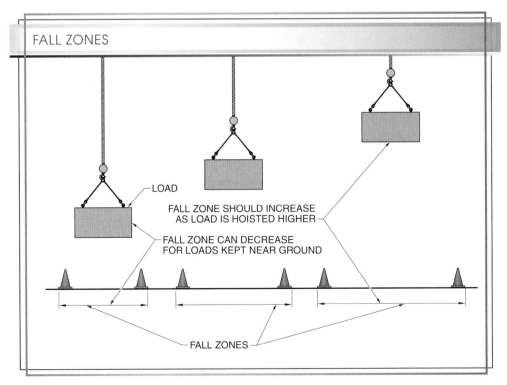

Figure 10-5. The fall zone is an area directly underneath a hoisted load where debris may fall if the rigging or hoist fails. As a load is hoisted higher, the fall zone should increase.

Note that obstacles include not just physical structures and equipment but also prohibited areas. For example, loads may not be hoisted over areas where personnel are present. **See Figure 10-6.** This makes the entire space above this area an obstacle that the load must not enter. Also, if the load's destination is within a structure, that presents additional obstacles. It may even be necessary to temporarily remove parts of the structure in order to place a large load.

OBSTRUCTIONS

Figure 10-6. Obstructions include not only physical equipment and structures that must be avoided by the crane and load but also prohibited areas that the load must not enter.

Load Preparation

The lift plan must consider any special preparation needed to make the load secure and ready for hoisting. This may require additional equipment and preparation time, which can affect the lift schedule. Therefore, the methods and equipment needed should be planned in advance.

Securing a Load. Some loads must be secured to prevent any movement or shifting. Loads consisting of a number of smaller components must be bound tightly into a single unit. Some loads, such as oddly shaped loads that are difficult to rig, require special equipment for securing and transporting. **See Figure 10-7.** Also, movable parts must be secured in a safe position.

Load binding straps, tarpaulins, or plastic stretch wrapping can be used to secure a load. The choice of securing equipment depends heavily on the type and configuration of the load. The securing equipment must not damage the load components but must be strong enough to hold under stress, such as if the load were to shift suddenly during the lift. Loads that cannot be adequately secured must be broken down into smaller portions to be moved safely.

Softeners. There is often contact between the load and the rigging components. Rigging often lays on top of the load while the hoisting line is slack. Also, rigging is often wrapped around

SPECIAL LOAD-SECURING EQUIPMENT

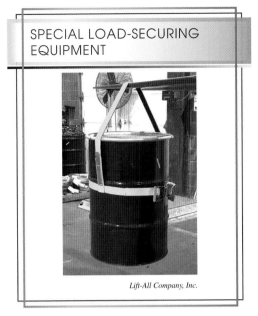

Lift-All Company, Inc.

Figure 10-7. Some loads, often due to having an awkward shape, require special equipment in order to be secured and attached to rigging.

loads as part of the attaching process. If the load is fragile or has a fine surface, it must be protected from the rigging components, which can be heavy, hard, or abrasive. Likewise, any sharp edges on a load can damage certain types of rigging, such as wire rope and woven fiber straps.

A *softener* is a relatively soft material used between loads and rigging in order to limit damaging contact. A number of materials can be used as softeners. Some riggers stock a selection of softeners of various shapes and sizes, or sometimes other materials on hand can be used as softeners. Common types of softeners include wood blocks, sections of pipe, fiber pads, scraps of carpet, and preformed plastic or metal corner protectors. **See Figure 10-8.** If the lift plan indicates that contact damage is a concern, then a selection of appropriate softeners should be part of the equipment list prepared prior to the lift.

SOFTENERS

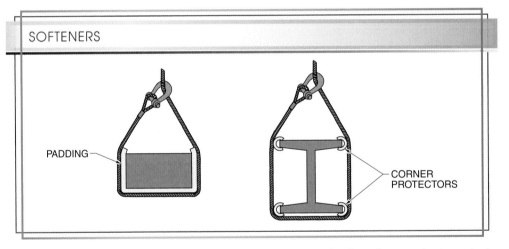

Figure 10-8. Softeners are used to protect the load and the rigging from damage due to contact with each other.

Cribbing. Cribbing is often used to support a load both before and after a lift. *Cribbing* is blocking used to temporarily support a load while at rest. **See Figure 10-9.** The cribbing materials and their arrangement must be strong enough to hold the load stable without deforming or collapsing. Wood blocks and beams are typically used for cribbing because they are strong but soft enough to not damage the load surface.

Some cribbing may consist simply of a short stack of blocks, while other cribbing is purposely built to support an oddly shaped load. Cribbing is also sometimes used to hold a load slightly off the ground so that rigging can be threaded underneath it. Loads may arrive at the job site already supported by cribbing. However, sometimes there is no cribbing, and blocks must be wedged beneath the load so rigging can be attached.

CRIBBING

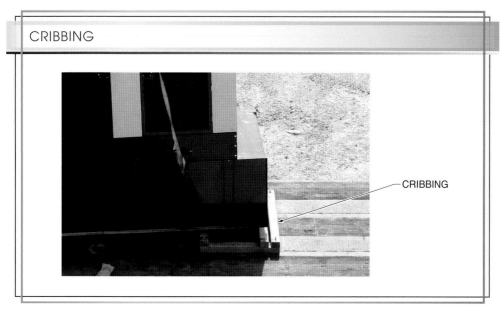

CRIBBING

Figure 10-9. Cribbing is blocking, usually wood blocks or planks, used to temporarily elevate, cradle, or secure a load.

Load Control

The lift plan should address how a load will be controlled as it is lifted. Load control involves maintaining both the stability and the orientation of a load.

An unstable load can cause catastrophic failures during a lift. Awkwardly shaped loads, loads consisting of bundled components, and loads with uneven weight distribution are particularly problematic. They can place unequal stresses on rigging components and are more prone to shifting unexpectedly. Maintaining control of such a load requires careful analysis of the load weight, center of gravity, and points of attachment. These items should be determined as early as possible in the planning process because they are the basis by which the rigging equipment is selected. In fact, some loads require special rigging equipment in order to be attached to a hoist hook.

The orientation of the load must be maintained during the entire lift, particularly during the landing phase. The load may need to face a certain direction in order to be installed in place or prepared to be moved horizontally to another location. When properly rigged, a load's orientation can be easily controlled with one or more tag lines. A *tag line* is a rope, handled by a qualified individual, used to control rotational movement of a load during a lift. **See Figure 10-10.** Personnel can easily rotate or hold a load by applying relatively light force on a tag line.

Loads placed on wheeled equipment for horizontal movement must also be controlled. These low-friction devices can cause a runaway load if the load is on any incline, even an incline that is not obvious. Manual hoists or block and tackle may be necessary to hold the load in place and control its descent down an incline. **See Figure 10-11.**

MATERIAL HANDLING

Many lifts involve some on-the-ground handling and horizontal movement of the load both before and after the lift. Ramps, floor openings, curbs, and obstructions may create load-handling problems that require special equipment. This material-handling phase may also present safety hazards that require consideration in a lift plan.

TAG LINES

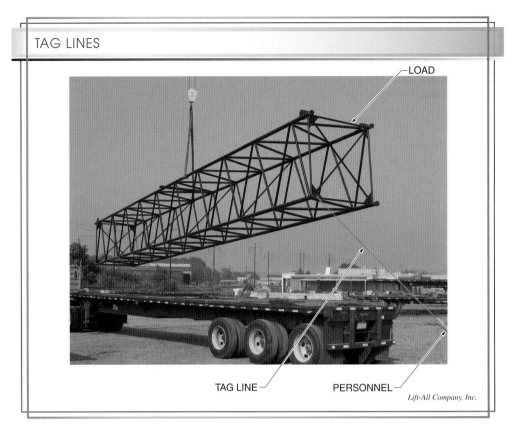

Lift-All Company, Inc.

Figure 10-10. Tag lines are used by personnel to help control the rotational movement of a hoisted load during a lift.

LOAD CONTROL ON AN INCLINE

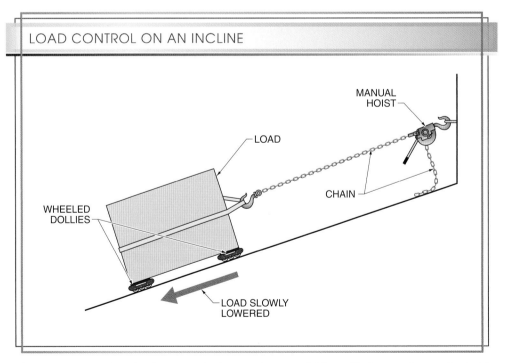

Figure 10-11. Loads on inclines must be kept under control. Manual hoists can be used to hold a load or slowly lower it down an incline.

Material handling equipment typically contacts and supports a load over a relatively small area, which is often concentrated at only a few points. **See Figure 10-12.** This increases the pressure and stress at these points. It must be determined whether the load or its supporting structure is strong and rigid enough to withstand this arrangement. Also, some equipment can only be supported at certain points, and these points will typically be indicated by the manufacturer on the load or in accompanying documentation.

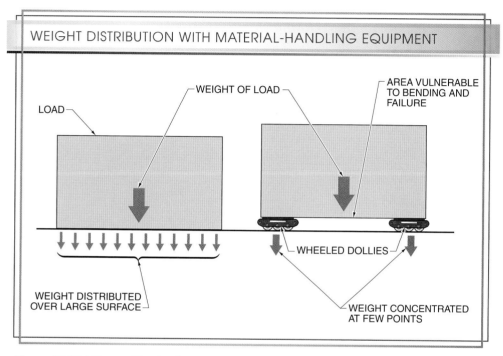

WEIGHT DISTRIBUTION WITH MATERIAL-HANDLING EQUIPMENT

Figure 10-12. When putting loads on material-handling equipment, such as dollies, the rigidity of the load and the stress of the concentrated weight must be considered.

Moving a load horizontally requires applying a pushing or pulling force to one side of the load. If the load is top-heavy or the force is applied high on the load, it may tip over. **See Figure 10-13.** A load should be examined for how its mass is distributed throughout its volume, particularly vertically. If the load was originally delivered on a pallet that was substantially larger than the base of the load, this is a clue that the load is top-heavy because a large pallet adds stability. It is usually best to apply horizontal forces at a low point on the load.

A number of different types of material-handling equipment may be involved in moving loads across floors or up and down inclines. Like all other rigging and hoisting equipment, material handling equipment must have a rated load that is sufficient to safely support the weight of their portion of a load.

Pry Bars and Lever Dollies

In most cases, a load must be lifted slightly in order to position material-handling equipment under it. A *pry bar,* also called a pinch bar, is a forged steel bar with an angled and flattened end that is pushed under the edge of a load in order to lift the load slightly. **See Figure 10-14.** The bar lifts the load as pressure is applied to the handle. Wedges, rollers, or other blocking can then be inserted under the load to allow space for other equipment.

APPLYING HORIZONTAL FORCES

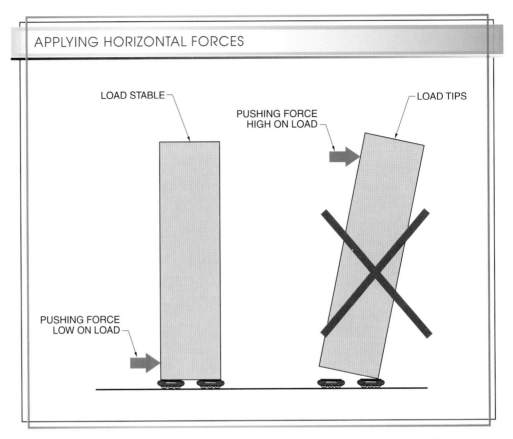

Figure 10-13. When moving loads horizontally, the force on the load must be applied at a low point. Pushing or pulling near the top of a load may cause tipping.

PRY BARS AND LEVER DOLLIES

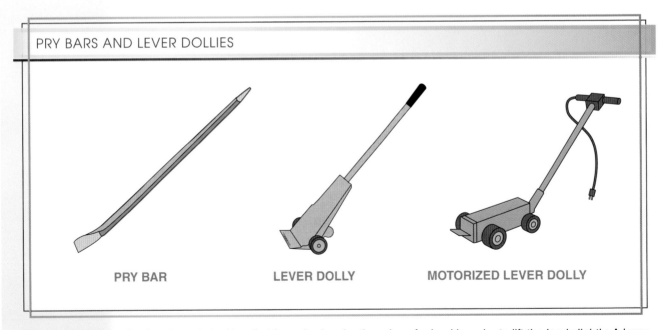

Figure 10-14. A pry bar is a forged steel bar that is pushed under the edge of a load in order to lift the load slightly. A lever dolly is a pry bar with wheels.

A *lever dolly,* also called a Johnson bar, is a pry bar with wheels that is used to raise a load slightly or reposition a barely hoisted load. The handle is very long, which provides a great amount of leverage force at the base. There is very little support under the load, so movements should be minimal, or the load could shift off of the lever dolly. These devices are generally reserved for making small adjustments in the position of a load on the floor or for placing other equipment under a lifted load. There are also motorized lever dollies, also known as electric Johnson bars. When using a motorized dolly, care should be taken to avoid tripping or rolling over the power cord.

Rollers

A flat-bottomed load can be moved a relatively short, horizontal distance across a hard, level, and smooth floor using simple rollers. Rollers are simply lengths of pipe or round steel stock that are placed under a load to provide a lower-friction surface. **See Figure 10-15.** As the load is moved forward, the rollers roll under the load, moving from the front to the back. As the load moves completely off a roller, the roller can be repositioned in front of the load to start again. Care should be taken so that the load does not overrun the rollers, causing it to tip forward. Using an ample number of rollers will fully support the load and prevent tipping.

If the rollers are arranged parallel to each other, the load will move in a straight line. By arranging the rollers in an arc, the load can also be turned as it moves.

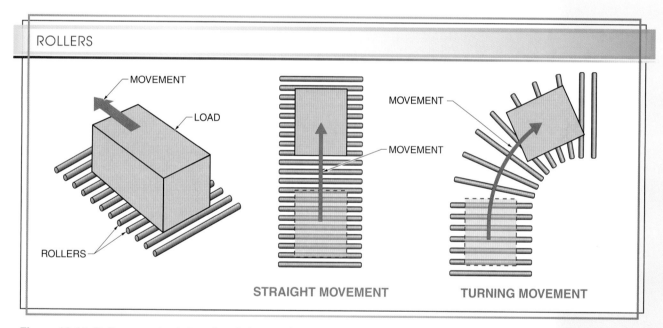

ROLLERS

STRAIGHT MOVEMENT

TURNING MOVEMENT

Figure 10-15. Rollers are simply lengths of pipe or other round stock that are used to move a load across a floor. By placing the rollers in an arc, the load can be steered slightly.

Air Casters

An *air caster* is a pneumatic device that is placed under a load to allow horizontal movement of the load. Air casters are connected to an air compressor, and the force of the compressed air under the air casters lifts the load slightly and allows a nearly frictionless movement of the load. **See Figure 10-16.** Multiple air casters can be connected together into palletlike or planklike structures to accommodate larger loads. Consideration should be given to the length and routing of the air supply line to prevent interference with the load-moving operation.

AIR CASTERS

AirFloat

Figure 10-16. Air casters use compressed air to create a nearly frictionless surface under a load.

Rigging Dollies

A *dolly* is a wheeled platform, sometimes with a handle, that is placed under a load to allow horizontal movement. Very simple dollies made from wood or metal frames and basic swivel casters may also be used with smaller loads.

A *rigging dolly* is a heavy-duty dolly, often with rollers, used to move large loads. Some rigging dollies are equipped with swivel pads to allow a load to pivot relative to the dolly. A removable handle is used to move and steer the load. **See Figure 10-17.** Load-equalizing dollies are special rigging dollies designed to balance the pressure of each wheel on the floor, which is important for nonuniform loads. This helps prevent damage to the floor as the equipment is moved.

RIGGING DOLLIES

Hilman Incorporated

Figure 10-17. Rigging dollies are heavy-duty wheels or rollers enclosed in a low-profile frame. These are placed under loads to transport them across smooth floors.

Machinery Lifts

A *machinery lift* is a two-wheeled dolly with swivel casters and a hydraulic lift, which is placed at the edge of a load. Machinery lifts are typically used in pairs, one on each side of a load, forming what is essentially a large four-wheeled dolly under the load. **See Figure 10-18.** A ratcheting strap is used to temporarily bind the two lifts to the load. The hydraulic lift is then used to raise the load slightly off of the floor, allowing the whole assembly to be rolled in any direction. Additional lifts can be employed in the middle of the load to allow it to be moved over short steps or curbs.

MACHINERY LIFTS

Skarnes, Inc.

Figure 10-18. A machinery lift is a combination dolly and hydraulic lift. Two or more machinery lifts are strapped around a load to lift and move it across a floor.

Pallet Jacks

A *pallet jack* is a dolly consisting of two forks that are hydraulically lifted above a set of wheels. When the forks are inserted into a standard pallet, the wheels roll into the spaces between the bottom slats and remain in contact with the floor. A *pallet* is an open structure, typically made of wood, that supports a load and provides openings for lifting equipment to be easily placed under the load. As the hydraulic jack is pumped with the handle, the wheels pivot down from inside the forks, raising the load. **See Figure 10-19.**

The load can then be moved via the wheels and guided by the handle. Some pallet jacks use an electric motor to move a load. The load is lowered by releasing the pressure in the hydraulic cylinder. The main advantage of a pallet jack is that as long as the load is on a pallet, the load does not need to be lifted to place a jack underneath it.

Forklifts

A *forklift* is a vehicle with hydraulically operated forks used to lift and transport loads. **See Figure 10-20.** A forklift is capable of placing loads at higher levels than a pallet jack but usually only up to about 10′ to 25′, much less than a crane. Forklifts are available in many combinations of engine type, tire type, reach height, and operator position. There are also many attachments available for lifting certain types of loads. When overloaded, a forklift is prone to tipping, so its rated load must be checked against the load weight. Also, the ground or floor that the forklift will operate on must be able to support the combined weight of the lift and the load.

Similar to crane operators, forklift operators must be properly trained and qualified. There are also hand signals that can be used by lifting personnel to guide the movement of a forklift.

PALLET JACKS

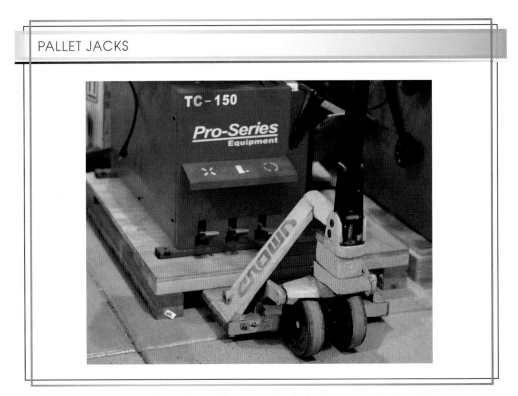

Figure 10-19. A pallet jack can be used to move loads that are on a sturdy pallet.

FORKLIFTS

FORKLIFT
TRUCK

FORKS
TILTED UP

Figure 10-20. A forklift is a vehicle that can lift loads, similar to a pallet jack. However, it can lift loads much higher than a pallet jack.

PRELIFT CHECKS

The lift-planning process should eventually lead to a well-prepared and trained crew with all the equipment and precautions necessary to complete a safe and effective lift. The final step in lift planning is a prelift check that confirms all expected conditions and dynamics. This is a "dry run" of the actual lift that often exposes additional issues that must be addressed before the actual lift.

All rigging equipment and hardware should be checked to ensure that all rated load specifications exceed the actual forces that each component will experience. This includes the crane or hoist and all under-hook lifting devices and hardware. It should be noted that depending on where the devices are located in the rigging, the required rated load may be different. Calculations for all expected forces should be double-checked. Also, it should be kept in mind that a hoist must also support the weight of all rigging components.

A test lift is made to confirm load balance and stability. During a test lift, the load is lifted only slightly off its support, just enough so that any issues can be identified. If the load is not level, then the rigging needs to be adjusted. In this case, the load must first be set back down on firm supports and the tension released from the rigging. This procedure is continued until the load is balanced and stable.

Any cribbing used to support a load before and after a lift must be checked for sufficient capacity and stability to support the load. The cribbing components should be staged in the areas where they will be needed prior to starting the lift. These staging areas may include intermediate locations where the load will be set down temporarily during the lifting operation.

Upon completion of the lift, the load may need to be moved to a new position or secured in place. Any rigging equipment that is no longer needed should be removed from the load and returned to storage. Often, the rigging equipment is left attached to the hook, and the crane is used to transport the equipment back to the storage area. Unused rigging equipment should not be allowed to accumulate in the work area where it may create a tripping hazard or become damaged.

COMMON APPLICATIONS— REPLACING UTILITY POLES

Utility poles are subjected to adverse conditions and may become damaged or break. To replace a broken utility pole, conductor wires are removed from the broken pole and held out of the way using proper safety procedures and cover-up. A new hole is dug or a utility pole is placed into the existing hole using the proper pole sling. The pole is then plumbed, and the hole is backfilled and tamped. The conductor wires are attached to the new utility pole and all hardware is transferred. The broken utility pole is removed in sections in most cases and loaded on the pole trailer.

Altec, Inc.

Chapter 10 Learner Resources

For additional information, visit
qr.njatcdb.org
Item # 1649

Appendix

Selected Hoisting-Related OSHA Regulations*

1910 Subpart N – Materials Handling and Storage
  1910.176 – Handling materials – general.
  1910.178 – Powered industrial trucks.
  1910.179 – Overhead and gantry cranes.
  1910.180 – Crawler locomotive and truck cranes.
  1910.181 – Derricks.
  1910.183 – Helicopters.
  1910.184 – Slings.
1910 Subpart R – Special Industries
  1910.269 – Electric Power Generation, Transmission, and Distribution
1926 Subpart H – Materials Handling, Storage, Use, and Disposal
  1926.251 – Rigging equipment for material handling.
1926 Subpart N – Helicopters, Hoists, Elevators, and Conveyors
  1926.551 – Helicopters.
  1926.552 – Material hoists, personnel hoists, and elevators.
  1926.553 – Base-mounted drum hoists.
  1926.554 – Overhead hoists.
  1926.555 – Conveyors.
  1926.556 – Aerial lifts.
1926 Subpart CC – Cranes & Derricks in Construction
  1926.1400 – Scope.
  1926.1401 – Definitions.
  1926.1402 – Ground conditions.
  1926.1403 – Assembly/Disassembly—selection of manufacturer or employer procedures.
  1926.1404 – Assembly/Disassembly—general requirements (applies to all assembly and disassembly operations).
  1926.1405 – Disassembly—additional requirements for dismantling of booms and jibs (applies to both the use of manufacturer procedures and employer procedures).
  1926.1406 – Assembly/Disassembly—employer procedures—general requirements.
  1926.1407 – Power line safety (up to 350 kV)—assembly and disassembly.
  1926.1408 – Power line safety (up to 350 kV)—equipment operations.
  1926.1409 – Power line safety (over 350 kV).
  1926.1410 – Power line safety (all voltages)—equipment operations closer than the Table A zone.
  1926.1411 – Power line safety—while traveling under or near power lines with no load.

1926.1412 – Inspections.
1926.1413 – Wire rope—inspection.
1926.1414 – Wire rope—selection and installation criteria.
1926.1415 – Safety devices.
1926.1416 – Operational aids.
1926.1417 – Operation.
1926.1418 – Authority to stop operation.
1926.1419 – Signals—general requirements.
1926.1420 – Signals—radio, telephone, or other electronic transmission of signals.
1926.1421 – Signals—voice signals—additional requirements.
1926.1422 – Signals—hand signal chart.
1926.1423 – Fall protection.
1926.1424 – Work area control.
1926.1425 – Keeping clear of the load.
1926.1426 – Free fall and controlled load lowering.
1926.1427 – Operator qualification and certification.
1926.1428 – Signal person qualifications.
1926.1429 – Qualifications of maintenance & repair employees.
1926.1430 – Training.
1926.1431 – Hoisting personnel.
1926.1432 – Multiple-crane/derrick lifts—supplemental requirements.
1926.1433 – Design, construction, and testing.
1926.1434 – Equipment modifications.
1926.1435 – Tower cranes.
1926.1436 – Derricks.
1926.1437 – Floating cranes/derricks and land cranes/derricks on barges.
1926.1438 – Overhead & gantry cranes.
1926.1439 – Dedicated pile drivers.
1926.1440 – Sideboom cranes.
1926.1441 – Equipment with a rated hoisting/lifting capacity of 2,000 pounds or less.
1926.1442 – Severability.
1926 Subpart CC App A – Standard Hand Signals
1926 Subpart CC App B – Assembly/Disassembly—Sample Procedures for Minimizing the Risk of Unintended Dangerous Boom Movement
1926 Subpart CC App C – Operator Certification—Written Examination—Technical Knowledge Criteria

*Does not include all applicable general construction or industry-specific regulations.

ASME B30 Standards

B30.1, *Jacks, Industrial Rollers, Air Casters, and Hydraulic Gantries*

B30.2, *Overhead and Gantry Cranes (Top Running Bridge, Single or Multiple Girder, Top Running Trolley Hoist)*

B30.3, *Tower Cranes*

B30.4, *Portal, Tower, and Pedestal Cranes*

B30.5, *Mobile and Locomotive Cranes*

B30.6, *Derricks*

B30.7, *Base Mounted Drum Hoists*

B30.8, *Floating Cranes and Floating Derricks*

B30.9, *Slings*

B30.10, *Hooks*

B30.11, *Monorails and Underhung Cranes*

B30.12, *Handling Loads Suspended from Rotorcraft*

B30.13, *Storage/Retrieval (S/R) Machines and Associated Equipment*

B30.14, *Side Boom Tractors*

B30.15, *Mobile Hydraulic Cranes (Withdrawn and Consolidated with B30.5)*

B30.16, *Overhead Hoists (Underhung)*

B30.17, *Overhead and Gantry Cranes (Top Running Bridge, Single Girder, Underhung Hoist)*

B30.18, *Stacker Cranes (Top or Under Running Bridge, Multiple Girder with Top or Under Running Trolley Hoist)*

B30.19, *Cableways*

B30.20, *Below-the-Hook Lifting Devices*

B30.21, *Manually Lever-Operated Hoists*

B30.22, *Articulating Boom Cranes*

B30.23, *Personnel Lifting Systems*

B30.24, *Container Cranes*

B30.25, *Scrap and Material Handlers*

B30.26, *Rigging Hardware*

B30.27, *Material Placement Systems*

B30.28, *Balance Lifting Units*

B30.29, *Self-Erecting Tower Cranes*

B30.30, *Ropes (Future Release)*

USER'S GUIDE TO LIFTING . . .

 **USER'S GUIDE LIFTING**

ASME VERSION (8/10)

RISK MANAGEMENT	TERMINOLOGY	FOR ADDITIONAL SUPPORT
DEFINITION	**WORKING LOAD LIMIT (WLL)**	

RISK MANAGEMENT

DEFINITION

COMPREHENSIVE SET OF ACTIONS THAT REDUCES THE RISK OF A PROBLEM, A FAILURE, AN ACCIDENT

ASME B30.9 REQUIRES THAT SLING USERS SHALL BE TRAINED IN THE SELECTION, INSPECTION, CAUTIONS TO PERSONNEL, EFFECTS OF ENVIRONMENT, AND RIGGING PRACTICES. SLING IDENTIFICATION IS REQUIRED ON ALL TYPES OF SLINGS

ASME B30.26 REQUIRES THAT RIGGING HARDWARE USERS SHALL BE TRAINED IN THE SELECTION, INSPECTION, CAUTIONS TO PERSONNEL, EFFECTS OF ENVIRONMENT, AND RIGGING PRACTICES. ALL RIGGING HARDWARE TO BE IDENTIFIED BY MANUFACTURER WITH NAME OR TRADEMARK OF MANUFACTURER.

REFER TO THE CROSBY GROUP CATALOG AND OTHER PRODUCT APPLICATION INFORMATION.

TERMINOLOGY

WORKING LOAD LIMIT (WLL)

THE MAXIMUM MASS OR FORCE WHICH THE PRODUCT IS AUTHORIZED TO SUPPORT IN A PARTICULAR SERVICE.

PROOF TEST

A TEST APPLIED TO A PRODUCT SOLELY TO DETERMINE INJURIOUS MATERIAL OR MANUFACTURING DEFECTS.

ULTIMATE STRENGTH

THE AVERAGE LOAD OR FORCE AT WHICH THE PRODUCT FAILS OR NO LONGER SUPPORTS THE LOAD.

DESIGN FACTOR

AN INDUSTRIAL TERM DENOTING A PRODUCT'S THEORETICAL RESERVE CAPABILITY; USUALLY COMPUTED BY DIVIDING THE CATALOG ULTIMATE LOAD BY THE WORKING LOAD LIMIT. GENERALLY EXPRESSED AS A RATIO, e.g. 5 TO 1.

 *Load Rated*®

FOR ADDITIONAL SUPPORT

the Crosby group,

P.O. Box 3128
Tulsa Oklahoma 74101
Phone: (918) 834-4611
Fax: (918) 832-0940
1-800-777-1555
Web:
www.thecrosbygroup.com
E-Mail:
crosbygroup@thecrosbygroup.com

BLOCKS & FITTINGS FOR WIRE ROPE & CHAIN

**CROSBY® FITTINGS
LEBUS® McKISSICK®
WESTERN NATIONAL**

THE BASIC RIGGING PLAN

PLAN EVERY LIFT, INCLUDE THE FOLLOWING QUESTIONS WITH THE QUESTIONS YOUR EXPERIENCE PROVIDES:

1. WHO IS RESPONSIBLE (COMPETENT) FOR THE RIGGING?
2. HAS COMMUNICATIONS BEEN ESTABLISHED?
3. IS THE RIGGING IN ACCEPTABLE CONDITION?
4. IS THE RIGGING APPROPRIATE FOR LIFTING?
5. DOES THE RIGGING HAVE PROPER IDENTIFICATION?
6. DOES ALL GEAR HAVE KNOWN WORKING LOAD LIMITS?
7. WHAT IS THE WEIGHT OF THE LOAD?
8. WHERE IS THE LOAD'S CENTER OF GRAVITY?
9. WHAT IS THE SLING ANGLE?
10. WILL THERE BE ANY SIDE OR ANGULAR LOADING?
11. ARE THE SLINGS PADDED AGAINST CORNERS, EDGES PROTRUSIONS AND ABRASIVE SURFACES?
12. ARE THE WORKING LOAD LIMITS ADEQUATE?
13. IS THE LOAD RIGGED TO THE CENTER OF GRAVITY?
14. IS THE HITCH APPROPRIATE FOR THE LOAD?
15. IS A TAG LINE REQUIRED TO CONTROL LOAD?
16. WILL PERSONNEL BE CLEAR OF SUSPENDED LOADS?
17. IS THERE ANY POSSIBILITY OF FOULING?
18. WILL THE LOAD LIFT LEVEL AND BE STABLE?
19. ANY UNUSUAL ENVIRONMENTAL CONCERNS?
20. ANY SPECIAL REQUIREMENTS?

THE RIGGING MUST BE USED WITHIN MANUFACTURER'S RECOMMENDATIONS AND INDUSTRY STANDARDS THAT INCLUDE OSHA, ASME, ANSI, API AND OTHERS.

RESPONSIBILITY

USER RESPONSIBILITY

1. UTILIZE APPROPRIATE RIGGING GEAR SUITABLE FOR OVERHEAD LIFTING.
2. UTILIZE THE RIGGING GEAR WITHIN INDUSTRY STANDARDS AND THE MANUFACTURER'S RECOMMENDATIONS.
3. CONDUCT REGULAR INSPECTION AND MAINTENANCE OF THE RIGGING GEAR.
4. PROVIDE EMPLOYEES WITH TRAINING TO MEET OSHA AND ASME (B30.9, B30.26, ETC.) REQUIREMENTS.

MANUFACTURERS RESPONSIBILITY

1. PRODUCT AND APPLICATION INFORMATION
2. PRODUCT THAT IS CLEARLY IDENTIFIED NAME OR LOGO LOAD RATING AND SIZE TRACEABILITY
3. PRODUCT PERFORMANCE WORKING LOAD LIMIT DUCTILITY FATIGUE PROPERTIES IMPACT PROPERTIES
4. PRODUCT TRAINING AND TRAINING RESOURCES

The Crosby Group LLC

INSPECTION OF RIGGING HARDWARE

INSPECTION FREQUENCY PER ASME B30.26

A VISUAL INSPECTION SHALL BE PERFORMED BY THE USER OR DESIGNATED PERSON EACH DAY BEFORE THE RIGGING HARDWARE IS USED.

A PERIODIC INSPECTION SHALL BE PERFORMED BY A DESIGNATED PERSON, AT LEAST ANNUALLY. THE RIGGING HARDWARE SHALL BE EXAMINED AND A DETERMINATION MADE AS TO WHETHER THEY CONSTITUTE A HAZARD. WRITTEN RECORDS ARE NOT REQUIRED.

SEMI-PERMANENT AND INACCESSIBLE LOCATIONS WHERE FREQUENT INSPECTIONS ARE NOT FEASIBLE SHALL HAVE PERIODIC INSPECTIONS PERFORMED.

REJECTION CRITERIA PER ASME B30.26

MISSING OR ILLEGIBLE MANUFACTURER'S NAME OR TRADEMARK AND/OR RATED LOAD IDENTIFICATION (OR SIZE AS REQUIRED)

A 10% OR MORE REDUCTION OF THE ORIGINAL DIMENSION

BENT, TWISTED, DISTORTED, STRETCHED, ELONGATED, CRACKED OR BROKEN LOAD BEARING COMPONENTS

EXCESSIVE NICKS, GOUGES, PITTING AND CORROSION

INDICATIONS OF HEAT DAMAGE INCLUDING WELD SPATTER OR ARC STRIKES, EVIDENCE OF UNAUTHORIZED WELDING

LOOSE OR MISSING NUTS, BOLTS, COTTER PINS, SNAP RINGS, OR OTHER FASTENERS AND RETAINING DEVICES

UNAUTHORIZED REPLACEMENT COMPONENTS OR OTHER VISIBLE CONDITIONS THAT CAUSE DOUBT AS TO THE CONTINUED USE OF THE SLING

ADDITIONALLY INSPECT WIRE ROPE CLIPS FOR:

1. INSUFFICIENT NUMBER OF CLIPS
2. INCORRECT SPACING BETWEEN CLIPS
3. IMPROPERLY TIGHTENED CLIPS
4. INDICATIONS OF DAMAGED WIRE ROPE OR WIRE ROPE SLIPPAGE
5. IMPROPER ASSEMBLY

ADDITIONALLY, INSPECT WEDGE SOCKETS FOR:

1. INDICATIONS OF DAMAGED WIRE ROPE OR WIRE ROPE SLIPPAGE
2. IMPROPER ASSEMBLY

ADDITIONAL REJECTION CRITERIA PER ASME B30.10 - HOOKS

ANY VISIBLY APPARENT BEND OR TWIST FROM THE PLANE OF THE UNBENT HOOK
ANY DISTORTION CAUSING AN INCREASE IN THROAT OPENING OF 5%, NOT TO EXCEED 1/4"

INSPECTION OF SLINGS

INSPECTION FREQUENCY PER ASME B30.9

A VISUAL INSPECTION FOR DAMAGE SHALL BE PERFORMED BY THE USER OR DESIGNATED PERSON EACH DAY OR SHIFT THE SLING IS USED. A COMPLETE INSPECTION FOR DAMAGE SHALL BE PERFORMED PERIODICALLY BY A DESIGNATED PERSON, AT LEAST ANNUALLY. WRITTEN RECORDS OF MOST RECENT PERIODIC INSPECTION SHALL BE MAINTAINED.

REJECTION CRITERIA PER ASME B30.9

MISSING OR ILLEGIBLE SLING IDENTIFICATION; EVIDENCE OF HEAT DAMAGE; SLINGS THAT ARE KNOTTED; FITTINGS THAT ARE PITTED, CORRODED, CRACKED, BENT, TWISTED, GOUGED, OR BROKEN; OTHER CONDITIONS, INCLUDING VISIBLE DAMAGE, THAT CAUSE DOUBT AS TO THE CONTINUED USE OF THE SLING.

WIRE ROPE SLINGS	CHAIN SLINGS	WEB SLINGS	ROUND SLINGS
EXCESSIVE BROKEN WIRES, FOR STRAND-LAID AND SINGLE PART SLINGS, TEN RANDOMLY DISTRIBUTED BROKEN WIRES IN ONE ROPE LAY OR FIVE BROKEN WIRES IN ONE STRAND IN ONE ROPE LAY	CRACKS OR BREAKS	ACID OR CAUSTIC BURNS	ACID OR CAUSTIC BURNS
	EXCESSIVE WEAR, NICKS OR GOUGES	MELTING OR CHARRING OF ANY PART OF THE SLING	EVIDENCE OF HEAT DAMAGE
	STRETCHED CHAIN LINKS OR COMPONENTS	HOLES, TEARS, CUTS OR SNAGS	HOLES, TEARS, CUTS, ABRASIVE WEAR OR SNAGS THAT EXPOSE THE CORE YARNS
SEVERE LOCALIZED ABRASION OR SCRAPING, KINKING, CRUSHING, BIRDCAGING	BENT, TWISTED OR DEFORMED CHAIN LINKS OR COMPONENTS	BROKEN OR WORN STITCHING IN LOAD BEARING SPLICES	BROKEN OR DAMAGED CORE YARNS
	EXCESSIVE PITTING OR CORROSION	EXCESSIVE ABRASIVE WEAR	WELD SPATTER THAT EXPOSES CORE YARNS
ANY OTHER DAMAGE RESULTING IN DAMAGE TO THE ROPE STRUCTURE	LACK OF ABILITY OF CHAIN OR COMPONENTS TO HINGE FREELY	DISCOLORATION AND BRITTLE OR STIFF AREAS ON ANY PART OF THE SLING, WHICH MAY MEAN CHEMICAL OR ULTRAVIOLET / SUNLIGHT DAMAGE	DISCOLORATION AND BRITTLE OR STIFF AREAS ON ANY PART OF THE SLINGS, WHICH MAY MEAN CHEMICAL OR OTHER DAMAGE
SEVERE CORROSION OF THE ROPE OR END ATTACHMENTS	WELD SPATTER		

The Crosby Group LLC

WIRE ROPE SLING CONNECTIONS AND HITCHES

CONNECTION TO FITTINGS

USE A THIMBLE TO PROTECT SLING AND INCREASE D/d

NEVER PLACE EYE OVER A FITTING SMALLER DIAMETER OR WIDTH THAN THE ROPE'S DIAMETER.

NEVER PLACE A SLING EYE OVER A FITTING WITH A DIAMETER OR WIDTH GREATER THAN ONE HALF THE LENGTH OF THE EYE.

CHOKER CAPACITY

A CHOKER HITCH HAS 75% OF THE CAPACITY OF A SINGLE LEG SLING ONLY IF THE ANGLE OF CHOKE IS 120 DEGREES OR GREATER. A CHOKE ANGLE LESS THAN 120 DEGREES CAN RESULT IN A CAPACITY AS LOW AS 40% OF THE SINGLE LEG.

BASKET HITCH CAPACITY

A BASKET HITCH HAS TWICE THE CAPACITY OF A SINGLE LEG ONLY IF D/d RATIO IS 25/1 AND THE LEGS ARE VERTICAL.

CAPACITY % OF	
ANGLE	SINGLE LEG
90	200 %
60	170 %
45	140 %
30	100 %

MULTIPLE LEG SLINGS

TRIPLE LEG SLINGS HAVE 50% MORE CAPACITY THAN DOUBLE LEG SLINGS (AT SAME SLING ANGLE) ONLY IF THE CENTER OF GRAVITY IS IN CENTER OF CONNECTION POINTS AND LEGS ARE ADJUSTED PROPERLY (THEY MUST HAVE AN EQUAL SHARE OF THE LOAD).

QUAD (4LEG) SLINGS OFFER IMPROVED STABILITY BUT PROVIDE INCREASED CAPACITY ONLY IF ALL LEGS SHARE AN EQUAL SHARE OF THE LOAD.

CHAIN SLING CONNECTIONS AND HITCHES

CONNECTION TO FITTINGS

USE MASTER LINKS TO COLLECT SLINGS AND TO CONNECT TO HOOK.

USE GRADE 8 (80) OR GRADE 10 (100) FITTINGS THAT MATCH THE W.L.L. OF CHAIN AND OFFER PROPER SECUREMENT.

CHOKER CAPACITY

A CHOKER HITCH HAS 80% OF THE CAPACITY OF A SINGLE LEG SLING ONLY IF THE ANGLE OF CHOKE IS 120 DEGREES OR GREATER. CHOKE ANGLES LESS THAN 120 DEGREES WILL RESULT IN A SIGNIFICANTLY REDUCED CAPACITY.

NO LOSS IN CAPACITY RESULTS IF A CRADLE GRAB HOOK IS USED.

BASKET HITCH CAPACITY

A TRUE BASKET HITCH HAS TWICE THE CAPACITY OF A SINGLE LEG ONLY IF THE LEGS ARE VERTICAL. NOTE THAT THE BASKET IS FORMED BY USING A CHAIN SLING WITH TWO MASTERLINKS AT EACH END CONNECTED TO THE HOOK.

HORIZONTAL CAPACITY %O F	
ANGLE	SINGLEL EG
90	200%
60	170%
45	140%
30	100%

MULTIPLE LEG SLINGS

TRIPLE LEG CHAIN SLINGS HAVE 50% MORE CAPACITY THAN DOUBLE LEG CHAIN SLINGS (AT SAME SLING ANGLE) ONLY IF THE CENTER OF GRAVITY IS IN CENTER OF CONNECTION POINTS AND LEGS ARE ADJUSTED PROPERLY (THEY MUST HAVE AN EQUAL SHARE OF THE LOAD). QUAD (4LEG) CHAIN SLINGS OFFER IMPROVED STABILITY, BUT DO NOT PROVIDE INCREASED CAPACITY. THE CAPACITY OF A FOUR LEG CHAIN SLING IS CONSIDERED THE SAME AS THREE LEG CHAIN SLING.

... USER'S GUIDE TO LIFTING ...

WEB SLING AND ROUNDSLING CAPACITIES

WEB SLING IDENTIFICATION INCLUDES:

SLING TYPE:
TC - TRIANGLE CHOKER,
TT - TRIANGLE TRIANGLE,
EE - EYE AND EYE,
EN - ENDLESS
    NUMBER OF PLIES: 1 OR 2
        WEBBING GRADE: 9 OR 6
           SLING WIDTH (INCH)

EE 2-9 04 X 12 ◄— SLING LENGTH (INCH)

ROUNDSLING IDENTIFICATION INCLUDES:

SLING NUMBER: 1-13
SLING NUMBERS ARE FOR REFERENCE
ONLY, SOME ROUNDSLINGS HAVE
DIFFERENT RATINGS.

SLING COLOR: PURPLE, GREEN, YELLOW,
TAN, RED, WHITE, BLUE, ORANGE
SLING COLOR IS NOT FOLLOWED BY ALL
MANUFACTURERS AND SOME COLORS HAVE
MORE THAN ONE RATED LOAD.

FOLDING, BUNCHING OR PINCHING
OF SYNTHETIC SLINGS, WHICH
OCCURS WHEN USED WITH
SHACKLES, HOOKS OR OTHER
APPLICATION WILL REDUCE THE
RATED LOAD.

CHOKER CAPACITY

A CHOKER HITCH HAS
80% OF THE CAPACITY
OF A SINGLE LEG SLING
ONLY IF THE ANGLE OF
CHOKE IS 120 DEGREES
OR GREATER. A CHOKE
ANGLE LESS THAN 120
DEGREES WILL
RESULT IN A
CAPACITY AS
LOW AS 40% OF
THE SINGLE LEG.

BASKET HITCH CAPACITY

HORIZONTAL CAPACITY % OF ANGLE SINGLE LEG	
90	200 %
60	170 %
45	140 %
30	100 %

A TRUE BASKET HITCH
HAS TWICE THE
CAPACITY OF A SINGLE
LEG ONLY IF THE LEGS
ARE VERTICAL.

MULTIPLE LEG SLINGS

TRIPLE LEG SLINGS HAVE 50% MORE
CAPACITY THAN DOUBLE LEG
SLINGS (AT SAME SLING ANGLE)
ONLY IF THE CENTER OF GRAVITY IS
IN CENTER OF CONNECTION POINTS
AND LEGS ADJUSTED PROPERLY
(THEY MUST HAVE AN EQUAL SHARE
OF THE LOAD).

QUAD (4LEG) SLINGS OFFER
IMPROVED STABILITY BUT PROVIDE
INCREASED CAPACITY ONLY IF ALL
LEGS SHARE AN EQUAL SHARE OF
THE LOAD.

ALWAYS SELECT AND USE WEB SLINGS AND ROUND SLINGS BY THE RATED LOAD SHOWN ON THE SLING IDENTIFICATION, NEVER BY WIDTH, COLOR OR SLING NUMBER.
NEVER PLACE A SYNTHETIC SLING EYE OVER A FITTING WITH A DIAMETER OR WIDTH GREATER THAN ONE THIRD THE LENGTH OF THE EYE.

CENTER OF GRAVITY AND SLING LOADING

WHEN LIFTING VERTICALLY,
THE LOAD WILL BE SHARED
EQUALLY IF THE CENTER OF
GRAVITY IS PLACED EQUALLY
BETWEEN THE PICK POINTS.
IF THE WEIGHT OF THE LOAD
IS 10,000 LBS., THEN EACH
SLING WILL HAVE A LOAD
OF 5,000 LBS. AND EACH
SHACKLE AND EYEBOLT WILL
ALSO HAVE A LOAD OF 5,000 LBS.

CENTER OF GRAVITY AND
SLING LOADING

SLING 1   SLING 2

10,000 LBS.

D1=5 FT   D2=5 FT

WEIGHTS AND MEASURES

UNIT WEIGHT STEEL = 490 LBS/FT³
UNIT WEIGHT ALUMINUM = 165 LBS/FT³
UNIT WEIGHT CONCRETE = 150 LBS/FT³
UNIT WEIGHT WOOD = 50 LBS/FT³
UNIT WEIGHT WATER = 62 LBS/FT³
UNIT WEIGHT SAND AND GRAVEL = 120 LBS/FT³
UNIT WEIGHT COPPER = 560 LBS/FT³
UNIT WEIGHT OIL = 58 LBS/FT³

1 CUBIC FT. = 7.5 GALS   1/2 INCH = 12.7 mm
1 METRIC TON = 1.1 US TONS   1 INCH = 25.4 mm
1 KILOGRAM = 2.2 LBS

CENTER OF GRAVITY AND SLING LOADING

WHEN THE CENTER OF GRAVITY
IS NOT EQUALLY SPACED BETWEEN
THE PICK POINTS, THE SLING AND
FITTINGS WILL NOT CARRY AN EQUAL
SHARE OF THE LOAD. THE SLING
CONNECTED TO THE PICK POINT
CLOSEST TO THE CENTER OF GRAVITY
WILL CARRY THE GREATEST SHARE
OF THE LOAD.

CENTER OF GRAVITY AND
SLING LOADING

SLING 1   SLING 2

10,000 LBS.

D1=8 FT   D2=2 FT

SLING 2 IS CLOSEST TO COG. IT WILL HAVE THE GREATEST SHARE
OF THE LOAD.
SLING 2 = 10,000 X 8 / (8+2) = 8,000 LBS.
SLING 1 = 10,000 X 2 / (8+2) = 2,000 LBS.

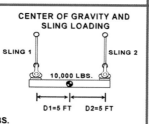

VOLUME OF CUBE =
HEIGHT x WIDTH x LENGTH

HEIGHT

WIDTH

VOLUME OF SPHERE =
3.14 x (DIAM. x DIAM. x DIAM.) / 6

VOLUME OF CYLINDER =
3.14 x (DIAM. x DIAM. x LENGTH) / 4

DIAMETER

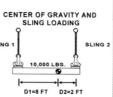

The Crosby Group LLC

. . . USER'S GUIDE TO LIFTING . . .

SLING ANGLES

TWO LEGGED SLING - WIRE ROPE, CHAIN, SYNTHETICS

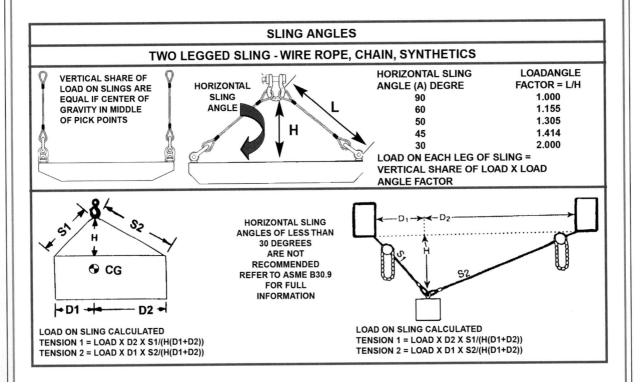

VERTICAL SHARE OF LOAD ON SLINGS ARE EQUAL IF CENTER OF GRAVITY IN MIDDLE OF PICK POINTS

HORIZONTAL SLING ANGLE

HORIZONTAL SLING ANGLE (A) DEGRE	LOADANGLE FACTOR = L/H
90	1.000
60	1.155
50	1.305
45	1.414
30	2.000

LOAD ON EACH LEG OF SLING = VERTICAL SHARE OF LOAD X LOAD ANGLE FACTOR

HORIZONTAL SLING ANGLES OF LESS THAN 30 DEGREES ARE NOT RECOMMENDED REFER TO ASME B30.9 FOR FULL INFORMATION

LOAD ON SLING CALCULATED
TENSION 1 = LOAD X D2 X S1/(H(D1+D2))
TENSION 2 = LOAD X D1 X S2/(H(D1+D2))

LOAD ON SLING CALCULATED
TENSION 1 = LOAD X D2 X S1/(H(D1+D2))
TENSION 2 = LOAD X D1 X S2/(H(D1+D2))

OPERATING PRACTICES - ASME B30.9

WHENEVER ANY SLING IS USED, THE FOLLOWING PRACTICES SHALL BE OBSERVED.

1. SLINGS THAT ARE DAMAGED OR DEFECTIVE SHALL NOT BE USED.
2. SLINGS SHALL NOT BE SHORTENED WITH KNOTS OR BOLTS OR OTHER MAKESHIFT DEVICES.
3. SLING LEGS SHALL NOT BE KINKED.
4. SLINGS SHALL NOT BE LOADED IN EXCESS OF THEIR RATED CAPACITIES.
5. SLINGS USED IN A BASKET HITCH SHALL HAVE THE LOADS BALANCED TO PREVENT SLIPPAGE.
6. SLINGS SHALL BE SECURELY ATTACHED TO THEIR LOAD.
7. SLINGS SHALL BE PADDED OR PROTECTED FROM THE SHARP EDGES OF THEIR LOADS.
8. SUSPENDED LOADS SHALL BE KEPT CLEAR OF ALL OBSTRUCTION.
9. ALL EMPLOYEES SHALL BE KEPT CLEAR OF LOADS ABOUT TO BE LIFTED AND OF SUSPENDED LOADS.
10. HANDS OR FINGERS SHALL NOT BE PLACED BETWEEN THE SLING AND ITS LOAD WHILE THE SLING IS BEING TIGHTENED AROUND THE LOAD.
11. SHOCK LOADING IS PROHIBITED!
12. A SLING SHALL NOT BE PULLED FROM UNDER A LOAD WHEN THE LOAD IS RESTING ON THE SLING.

INSPECTION: EACH DAY BEFORE BEING USED, THE SLING AND ALL FASTENINGS AND ATTACHMENTS SHALL BE INSPECTED FOR DAMAGE OR DEFECTS BY A COMPETENT PERSON DESIGNATED BY THE EMPLOYER. ADDITIONAL INSPECTIONS SHALL BE PERFORMED DURING SLING USE WHERE SERVICE CONDITIONS WARRANT. DAMAGED OR DEFECTIVE SLINGS SHALL BE IMMEDIATELY REMOVED FROM SERVICE.

LOAD CONTROL

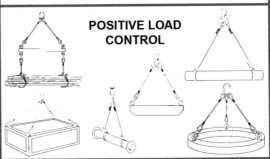

POSITIVE LOAD CONTROL

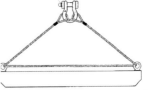

REEVING THROUGH CONNECTIONS TO LOAD INCREASES LOAD ON CONNECTION FITTINGS BY AS MUCH AS TWICE.

DO NOT REEVE!

The Crosby Group LLC

BLOCK SELECTION AND APPLICATION GUIDE . . .

VERSION (10/11/10) **Crosby** BLOCK SELECTION AND APPLICATION GUIDE 1

RISK MANAGEMENT

COMPREHENSIVE SET OF ACTIONS THAT REDUCES THE RISK OF A PROBLEM, A FAILURE, AN ACCIDENT

YOU NEED

- PRODUCT KNOWLEDGE
- APPLICATION KNOWLEDGE
- MANUFACTURER OF KNOWN CAPABILITY
- PRODUCTS THAT ARE CLEARLY IDENTIFIED WITH THE FOLLOWING:
  1. MANUFACTURER'S NAME AND LOGO
  2. LOAD RATING OR SIZE THAT REFERENCES RATINGS
  3. TRACEABILITY CODE

A GOOD RISK MANAGEMENT PROGRAM RECOGNIZES

- PERFORMANCE REQUIREMENTS INCLUDE THE FOLLOW:

1. LOAD RATED PRODUCTS
2. QUENCHED AND TEMPERED
3. ABILITY TO DEFORM WHEN OVERLOADED.
4. ABILITY TO WITHSTAND REAL WORLD LOADING IN DAY TO DAY USE, TOUGHNESS.

MECHANICAL ADVANTAGE AND TOTAL LOAD

MECHANICAL ADVANTAGE IS THE LEVERAGE GAINED BY A MULTIPLE PART BLOCK. MUST HAVE A TRAVELING BLOCK TO HAVE MECHANICAL ADVANTAGE. THE THEORETICAL ADVANTAGE IS EQUAL TO THE NUMBER OF PARTS OF LINE SUPPORTING THE TRAVELING BLOCK.

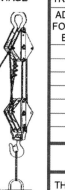

P

6000 lbs

TRUE MECHANICAL ADVANTAGE		
ADVANTAGE FOR BRONZE BUSHING	ADVANTAGE FOR ANTI FRICTION	NUMBER OF LINE PARTS
5.16	5.60	6
5.90	6.47	7
6.60	7.32	8
7.27	8.16	9
7.91	8.98	10
8.52	9.79	11
9.11	10.60	12

TOTAL LOAD

THE TOTAL LOAD PLACED ON THE BLOCK AND ITS END FITTING DETERMINES THE WORKING LOAD LIMIT REQUIRED.

theCrosbygroup

P.O.BOX 3128 TULSA, OK USA
(918) 834-4611   FAX (918) 832-0940
WWW.THECROSBYGROUP.COM

WORKING WITH BLOCKS   2

OVERHAUL WEIGHT

To determine the weight of the block or overhaul ball that is required to free fall the block, the following information is needed: **Size of wire rope, Number of line parts, Type of sheave bearing, Length of crane boom, and Drum Friction.**

BLOCK REEVING

Straight laced reeving is a basic method of placing the rope through a set of blocks. The end of the rope is fed through the outside sheave of the upper block to the outside sheave of the lower (traveling) block. This continues to the last sheave.

ADVANTAGES:
1. Allows blocks to run closer together
2. Is simple.
3. Has no reverse bends.

DRAWBACKS:
Tilting because of imbalanced loading can cause block rotation and wear of the sheaves and wire rope

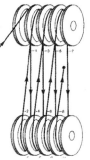

SYMMETRICAL REEVING

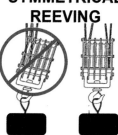

Reeve blocks symmetrically to distribute load evenly. All sheaves must be reeved to achieve the full working load limit of the block.

BLOCK CABLING

1. Reduce wire rope length
2. Use even part reeving
3. Dead end to boom
4. Evaluate wire rope construction

FOR ADDITIONAL INFORMATION REFER TO THE CROSBY GENERAL CATALOG

The Crosby Group LLC

... BLOCK SELECTION AND APPLICATION GUIDE ...

RIGGING WITH BLOCKS 3

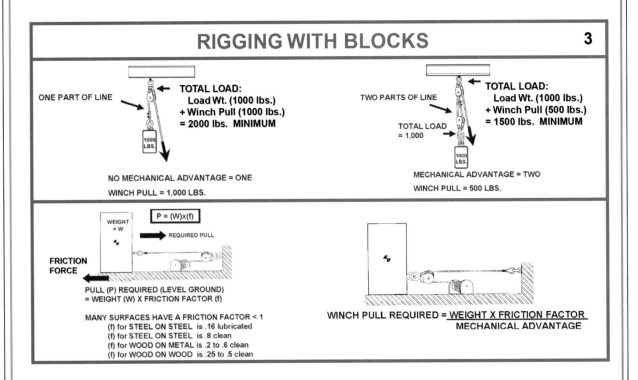

ONE PART OF LINE

TOTAL LOAD:
Load Wt. (1000 lbs.)
+ Winch Pull (1000 lbs.)
= 2000 lbs. MINIMUM

NO MECHANICAL ADVANTAGE = ONE
WINCH PULL = 1,000 LBS.

TWO PARTS OF LINE

TOTAL LOAD:
Load Wt. (1000 lbs.)
+ Winch Pull (500 lbs.)
= 1500 lbs. MINIMUM

TOTAL LOAD = 1,000

MECHANICAL ADVANTAGE = TWO
WINCH PULL = 500 LBS.

WEIGHT = W

$$P = (W)x(f)$$

REQUIRED PULL

FRICTION FORCE

PULL (P) REQUIRED (LEVEL GROUND)
= WEIGHT (W) X FRICTION FACTOR (f)

MANY SURFACES HAVE A FRICTION FACTOR < 1
(f) for STEEL ON STEEL is .16 lubricated
(f) for STEEL ON STEEL is .8 clean
(f) for WOOD ON METAL is .2 to .6 clean
(f) for WOOD ON WOOD is .25 to .5 clean

WINCH PULL REQUIRED = WEIGHT X FRICTION FACTOR / MECHANICAL ADVANTAGE

BLOCK LOADING - ANGLE FACTOR MULTIPLIERS 4

A single line sheave block used to change load line direction can be subject to total loads greatly different from the line pull

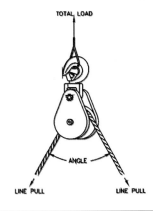

TOTAL LOAD

ANGLE

LINE PULL   LINE PULL

ANGLE FACTOR MULTIPLIERS

ANGLE	FACTOR	ANGLE	FACTOR
0°	2.00	100°	1.29
10°	1.99	110°	1.15
20°	1.97	120°	1.00
30°	1.93	130°	.84
40°	1.87	135°	.76
45°	1.84	140°	.68
50°	1.81	150°	.52
60°	1.73	160°	.35
70°	1.64	170°	.17
80°	1.53	180°	.00
90°	1.41	—	—

TOTAL LOAD = LINE PULL X ANGLE FACTOR
EXAMPLE, AT 45 DEGREES, AND 10,000 LB LINE PULL,
TOTAL LOAD = 10,000 X 1.84 = 18,400 LBS.

FOR ADDITIONAL INFORMATION REFER TO THE CROSBY GENERAL CATALOG

The Crosby Group LLC

SHEAVE INSPECTION 9

SHEAVE INSPECTION

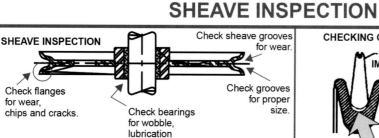

Check flanges for wear, chips and cracks.

Check bearings for wobble, lubrication & ease of rotation.

Check sheave grooves for wear.

Check grooves for proper size.

CHECKING GROOVE SIZE FOR PROPER SIZE

IMPROPER   PROPER

SHEAVE INSPECTION
Minimum groove radii for worn sheave
tolerances per "Wire Rope User's Manual" (third edition)

NOMINAL WIRE ROPE SIZE (in.)	RADII (in.)	NOMINAL WIRE ROPE SIZE (in.)	RADII (in.)
1/4	.128	3/4	.384
5/16	.160	7/8	.448
3/8	.192	1	.513
7/16	.224	1-1/8	.577
1/2	.256	1-1/4	.641
9/16	.266	1-3/8	.705
5/8	.320	1-1/2	.769

SHEAVE FLEET ANGLE*

● Fleet Angle is the entrance and exit angle of the wire rope relative to the sheave

● Fleet angle should be no more then 1-1/2 degrees

FLEET ANGLE

* NOTE: "Wire Rope User's Manual" allows 2 degrees on grooved winch drums.

FOR ADDITIONAL INFORMATION REFER TO THE CROSBY GENERAL CATALOG

BLOCK HOOK INSPECTION 10

CROSBY RECOMMENDS AS A MINIMUM:

1. A visual inspection for cracks, nicks, wear, gouges and deformation as part of a comprehensive documented inspection program, should be conducted by trained personnel in compliance with the schedule in ASME B30.10.

2. For hooks used in frequent load cycles or pulsating load, or exposed to corrosive conditions (Road Salt, etc.) the hook and thread should be periodically inspected by Magnetic Particle or Dye Penetrant.

LUBRICATION OF HOOK BEARINGS:

Anti Friction — Every 14 days for frequent swiveling; every 45 days for infrequent swiveling.

Bronze Thrust Bushing or No Bearing — Every 16 hours for frequent swiveling; every 21 days for infrequent swiveling.

ASME B30.10 INSPECTION FREQUENCY

1. **Frequent Inspection** - visual examinations by the operator or other designated person.
   (a) normal service - monthly.
   (b) heavy service - weekly to monthly
   (c) severe service - daily or weekly
   (d) special or infrequent service as authorized by a qualified person - before and after each occurrence, with records of the operation.

2. **Period Inspection** - visual inspections by an appointed person making records of apparent external conditions to provide the basis for continuing evaluation.
   (a) normal service - equipment in place - yearly;(definition: service, normal - service that involves operating at less than 85 percent of rated load except for isolated instances.)
   (b) heavy service - as in normal service, unless external conditions indicate that disassembly should be done to permit detailed inspection - yearly;(definition: service, heavy service that involves operating 85 percent to 100 percent of rated as a regular specified procedure.)
   (c) severe service - as in heavy service, except that the detailed inspection may show the need for use of nondestructive type of testing - quarterly;(definition: service, severe - heavy service coupled with abnormal operating conditions.)

FOR ADDITIONAL INFORMATION REFER TO ASME B30.10 AND OSHA 1910.179 OVERHEAD GANTRY CRANES

OSHA HAND SIGNALS

EMERGENCY STOP	STOP	DOG EVERYTHING
USE MAIN HOIST	USE AUXILIARY HOIST	HOIST	LOWER
MOVE SLOWLY	SWING	RAISE BOOM	LOWER BOOM
RAISE THE BOOM AND LOWER THE LOAD	LOWER THE BOOM AND RAISE THE LOAD	EXTEND TELESCOPING BOOM	RETRACT TELESCOPING BOOM
TRAVEL/ TOWER TRAVEL	TROLLEY TRAVEL	CRAWLER CRANE, BOTH TRACKS	CRAWLER CRANE TRAVEL, ONE TRACK

OSHA HELICOPTER HAND SIGNALS

MOVE RIGHT

HOLD HOVER

MOVE LEFT

TAKE OFF

MOVE FORWARD

LAND

MOVE REARWARD

MOVE UPWARD

RELEASE SLING LOAD

MOVE DOWNWARD

COMMON U.S. CUSTOMARY UNITS

	Unit	Abbreviation	Equivalents
Length	inch	in. or "	0.083 ft or 0.028 yd
	foot	ft or '	12 in. or 0.33 yd
	yard	yd	36 in. or 3 ft
Area	square inch	sq in. or in^2	0.0069 sq ft or 0.00077 sq yd
	square foot	sq ft or ft^2	144 sq in. or 0.11 sq yd
	square yard	sq yd or yd^2	1296 sq in. or 9 sq ft
Volume	cubic inch	cu in. or in^3	0.00058 cu ft or 0.000021 cu yd
	cubic foot	cu ft or ft^3	1728 cu in. or 0.037 cu yd
	cubic yard	cu yd or yd^3	46,659 cu in. or 27 cu ft
Weight	pound	lb or #	0.0005 t
	ton*	t	2000 lb

* known as a short ton outside of the United States

COMMON METRIC UNITS

	Unit	Abbreviation	Equivalents
Length	centimeter	cm	0.01 m or 0.00001 km
	meter	m	100 cm or 0.001 km
	kilometer	km	100,000 cm or 1000 m
Area	square centimeter	cm^2	0.0001 m^2
	square meter	m^2	10,000 cm^2
Volume	cubic centimeter	cm^3	0.000001 m^3 or 0.001 l
	cubic meter	m^3	1,000,000 cm^3 or 1000 l
	liter	l	1000 cm^3 or 0.001 m^3
Weight	kilogram	kg	0.001 t
	metric ton	t	1000 kg

UNIT CONVERSIONS

	U.S. Customary to Metric	Metric to U.S. Customary
Length	1 ft = 0.305 m	1 m = 3.28 ft
Area	1 ft^2 = 0.0929 m^2	1 m^2 = 10.8 ft^2
Volume	1 ft^3 = 0.0283 m^3	1 m^3 = 35.3 ft^3
Weight and Mass	1 lb = 0.454 kg	1 kg = 2.20 lb

METRIC PREFIXES

Multiples and Submultiples	Prefixes	Symbols	Meaning
1,000,000,000,000 = 10^{12}	tera	T	trillion
1,000,000,000 = 10^9	giga	G	billion
1,000,000 = 10^6	mega	M	million
1,000 = 10^3	kilo	k	thousand
100 = 10^2	hecto	h	hundred
10 = 10^1	deka	da	ten
1 = 10^0			
0.1 = 10^{-1}	deci	d	tenth
0.01 = 10^{-2}	centi	c	hundredth
0.001 = 10^{-3}	milli	m	thousandth
0.000001 = 10^{-6}	micro	μ	millionth
0.000000001 = 10^{-9}	nano	n	billionth
0.000000000001 = 10^{-12}	pico	p	trillionth

STANDARD STOCK MATERIAL SHAPES . . .

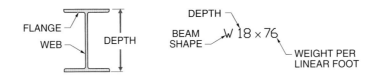

BEAM SHAPE SYMBOLS:

W (WIDE FLANGE), S (STANDARD), M (MISCELLANEOUS), HP (H-PILE)

BEAM

CHANNEL SHAPE SYMBOLS:

$\complement$, C (AMERICAN STANDARD), MC (MISCELLANEOUS CHANNEL)

CHANNEL

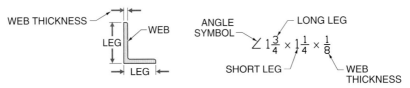

NOTE: LONG LEG DIMENSION LISTED FIRST.

ANGLE SYMBOLS: L, $\angle$

ANGLE

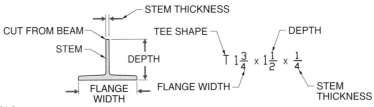

ANGLE SHAPE SYMBOLS:

T, WT (TEE CUT FROM W BEAM), ST (TEE CUT FROM S BEAM), MT (TEE CUT FROM M BEAM)

TEE

... STANDARD STOCK MATERIAL SHAPES ...

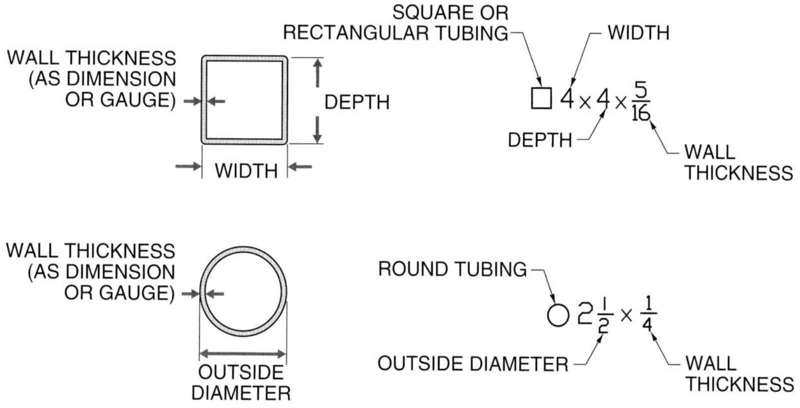

WALL THICKNESS
(AS DIMENSION
OR GAUGE)

SQUARE OR
RECTANGULAR TUBING

WIDTH

DEPTH

WIDTH

$\square\ 4 \times 4 \times \frac{5}{16}$

DEPTH

WALL
THICKNESS

WALL THICKNESS
(AS DIMENSION
OR GAUGE)

OUTSIDE
DIAMETER

ROUND TUBING

$\bigcirc\ 2\frac{1}{2} \times \frac{1}{4}$

OUTSIDE DIAMETER

WALL
THICKNESS

TUBING SHAPE SYMBOLS:
$\square$ (SQUARE OR RECTANGULAR), $\bigcirc$ (ROUND), TS (STRUCTURAL TUBING)

TUBING

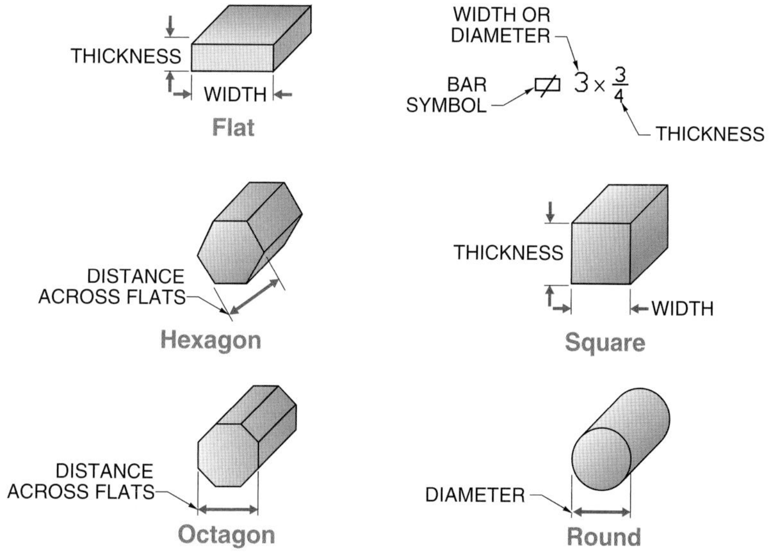

THICKNESS

WIDTH

Flat

WIDTH OR
DIAMETER

BAR
SYMBOL

$3 \times \frac{3}{4}$

THICKNESS

DISTANCE
ACROSS FLATS

Hexagon

THICKNESS

WIDTH

Square

DISTANCE
ACROSS FLATS

Octagon

DIAMETER

Round

BAR SYMBOLS:
BAR, $\varnothing$ (ROUND)

BAR

. . . STANDARD STOCK MATERIAL SHAPES

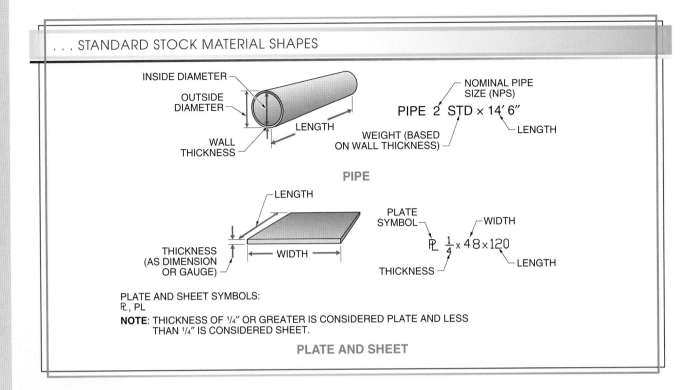

PIPE

PLATE AND SHEET SYMBOLS:
℞, PL

NOTE: THICKNESS OF ¼″ OR GREATER IS CONSIDERED PLATE AND LESS THAN ¼″ IS CONSIDERED SHEET.

PLATE AND SHEET

ELECTRICAL CONDUIT WEIGHTS*

Trade Size	Steel EMT	Steel IMC	Steel Rigid	Aluminum Rigid	PVC Rigid
½	0.300	0.620	0.820	0.281	0.164
¾	0.460	0.840	1.090	0.374	0.218
1	0.670	1.19	1.610	0.545	0.321
1¼	1.01	1.58	2.18	0.716	0.434
1½	1.16	1.94	2.63	0.887	0.518
2	1.48	2.56	3.50	1.19	0.695
2½	2.16	4.41	5.59	1.88	1.10
3	2.63	5.43	7.27	2.46	1.44
3½	3.49	6.29	8.80	2.96	1.73
4	3.93	7.00	10.3	3.50	2.04
5	—	—	14.0	4.79	2.78
6	—	—	18.4	6.30	3.60

* in lb/ft

STOCK MATERIAL WEIGHTS*

Material	Weight	Material	Weight	Material	Weight
Material		Chestnut	30	Granite	172
Aluminum, cast hammered	165	Cypress, southern	32	Greenstone, trap	187
Aluminum, bronze	481	Douglas fir	34	Gypsum, alabaster	159
Antimony	416	Elm, American	35	Limestone	160
Arsenic	358	Hemlock, eastern or western	28	Magnesite	187
Bismuth	608	Hickory	53	Marble	168
Brass, cast-rolled	534	Larch, western	36	Phosphate rock, apatite	200
Chromium	428	Maple, red or black	38 – 40	Pumice, natural	40
Cobalt	552	Oak	51	Quartz, flint	165
Copper, cast-rolled	556	Pine	27 – 28	Sandstone, bluestone	147
Gold, cast-hammered	1205	Poplar, yellow	28	Slate, shale	172
Iron, cast pig	450	Redwood	30	Soapstone, talc	169
Iron, wrought	485	Spruce	28	**Bituminous Substances**	
Iron, slag	172	Tamarack	37	Asphaltum	81
Lead	706	Walnut	39 – 40	Coal, anthracite	97
Magnesium	109	**Liquids**		Coal, bituminous	84
Manganese	456	Alcohol, 100%	49	Coal, lignite	78
Mercury	848	Acid, muriatic (40%)	75	Coal, coke	75
Molybdenum	562	Acid, nitric (91%)	94	Graphite	131
Nickel	545	Acid, sulphuric (87%)	112	Paraffin	56
Platinum, cast-hammered	1330	Lye, soda (66%)	106	Petroleum, crude	55
Silver, cast-hammered	656	Oils	58	Petroleum, refined	50
Steel	490	Petroleum	55	Pitch	69
Tin, cast-hammered	459	Gasoline	42	Tar, bituminous	75
Tungsten	1180	Water, at 4°C	62	**Brick Masonry**	
Vanadium	350	Water, ice	56	Pressed brick	140
Zinc, cast-rolled	440	Water, fresh snow	8	Common brick	120
Other Solids		Water, sea water	64	Soft brick	100
Carbon, amorphous	129	**Gases**			
Cork	15	Air, at 0°C	0.08071	Cement, stone, sand	144
Ebony	76	Ammonia	0.0478	Cement, slag, etc.	130
Fats	58	Carbon dioxide	0.1234	Cement, cinder, etc.	100
Glass, common plate	160	Carbon monoxide	0.0781	**Building Material**	
Glass, crystal	184	Gas, natural	0.038 – 0.039	Ashes, cinders	40 – 45
Phosphorous, white	114	Hydrogen	0.00559	Cement, Portland (loose)	90
Resins, rosin or amber	67	Nitrogen	0.0784	Cement, Portland (set)	183
Rubber	58	Oxygen	0.0892	Lime, gypsum (loose)	65 – 75
Silicon	155	**Minerals**		Mortar, set	103
Sulphur, amorphous	128	Asbestos	153	Slags, bank slag	67 – 72
Wax	60	Basalt	184	Slags, screenings	98 – 117
Timber, U.S. Seasoned		Bauxite	159	Slags, machine slag	96
Ash, white	41	Borax	109	**Earth**	
Beech	44	Chalk	137	Clay, damp	110
Birch, yellow	43	Clay	137	Dry, packed	95
Cedar, white or red	22 – 23	Dolomite	181	Mud, packed	115

* in lb/cu ft

REGULAR-NUT EYEBOLTS

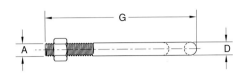
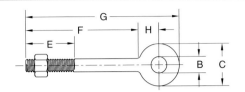

Shank Diameter and Length*	G-291 Stock No. Galv	Working Load Limit†	Weight per 100†	Dimensions*							
				A	B	C	D	E	F	G	H
¼ × 2	1043230	650	6.00	0.25	0.50	1.00	0.25	1.50	2.00	3.06	0.56
¼ × 4	1043258	650	13.50	0.25	0.50	1.00	0.25	2.50	4.00	5.06	0.56
⁵⁄₁₆ × 2¼	1043276	1200	18.75	0.31	0.62	1.25	0.31	1.50	2.25	3.56	0.69
⁵⁄₁₆ × 4¼	1043294	1200	25.00	0.31	0.62	1.25	0.31	2.50	4.25	5.56	0.69
⅜ × 2½	1043310	1550	24.33	0.38	0.75	1.50	0.38	1.50	2.50	4.12	0.88
⅜ × 4½	1043338	1550	37.50	0.38	0.75	1.50	0.38	2.50	4.50	6.12	0.88
⅜ × 6	1043356	1550	43.75	0.38	0.75	1.50	0.38	2.50	6.00	7.62	0.88
½ × 3¼	1043374	2600	50.00	0.50	1.00	2.00	0.50	1.50	3.25	5.38	1.12
½ × 6	1043392	2600	62.50	0.50	1.00	2.00	0.50	3.00	6.00	8.12	1.12
½ × 8	1043418	2600	75.00	0.50	1.00	2.00	0.50	3.00	8.00	10.12	1.12
½ × 10	1043436	2600	88.00	0.50	1.00	2.00	0.50	3.00	10.00	12.12	1.12
½ × 12	1043454	2600	100.00	0.50	1.00	2.00	0.50	3.00	12.00	14.12	1.12
⅝ × 4	1043472	5200	101.25	0.62	1.25	2.50	0.62	2.00	4.00	6.69	1.44
⅝ × 6	1043490	5200	120.00	0.62	1.25	2.50	0.62	3.00	6.00	8.69	1.44
⅝ × 8	1043515	5200	131.00	0.62	1.25	2.50	0.62	3.00	8.00	10.69	1.44
⅝ × 10	1043533	5200	162.50	0.62	1.25	2.50	0.62	3.00	10.00	12.69	1.44
⅝ × 12	1043551	5200	175.00	0.62	1.25	2.50	0.62	4.00	12.00	14.69	1.44
¾ × 4½	1043579	7200	185.90	0.75	1.50	3.00	0.75	2.00	4.50	7.69	1.69
¾ × 6	1043597	7200	180.00	0.75	1.50	3.00	0.75	3.00	6.00	9.19	1.69
¾ × 8	1043613	7200	200.00	0.75	1.50	3.00	0.75	3.00	8.00	11.19	1.69
¾ × 10	1043631	7200	237.50	0.75	1.50	3.00	0.75	3.00	10.00	13.19	1.69
¾ × 12	1043659	7200	251.94	0.75	1.50	3.00	0.75	4.00	12.00	15.19	1.69
¾ × 15	1043677	7200	300.00	0.75	1.50	3.00	0.75	5.00	15.00	18.19	1.69
⅞ × 5	1043695	10,600	275.00	0.88	1.75	3.50	0.88	2.50	5.00	8.75	2.00
⅞ × 8	1043711	10,600	325.00	0.88	1.75	3.50	0.88	4.00	8.00	11.75	2.00
⅞ × 12	1043739	10,600	400.00	0.88	1.75	3.50	0.88	4.00	12.00	15.75	2.00
1 × 5	1043757	13,300	425.00	1.00	2.00	4.00	1.00	3.00	6.00	10.31	2.31
1 × 9	1043775	13,300	452.00	1.00	2.00	4.00	1.00	4.00	9.00	13.31	2.31
1 × 12	1043793	13,300	550.00	1.00	2.00	4.00	1.00	4.00	12.00	16.31	2.31
1 × 18	1043819	13,300	650.00	1.00	2.00	4.00	1.00	7.00	18.00	22.31	2.31
1¼ × 8	1043837	21,000	750.00	1.25	2.50	5.00	1.25	4.00	8.00	13.38	2.88
1¼ × 12	1043855	21,000	900.00	1.25	2.50	5.00	1.25	4.00	12.00	17.38	2.88
1¼ × 20	1043873	21,000	1150.00	1.25	2.50	5.00	1.25	6.00	20.00	25.38	2.88

* in in.
† in lb

SHOULDER-NUT EYEBOLTS

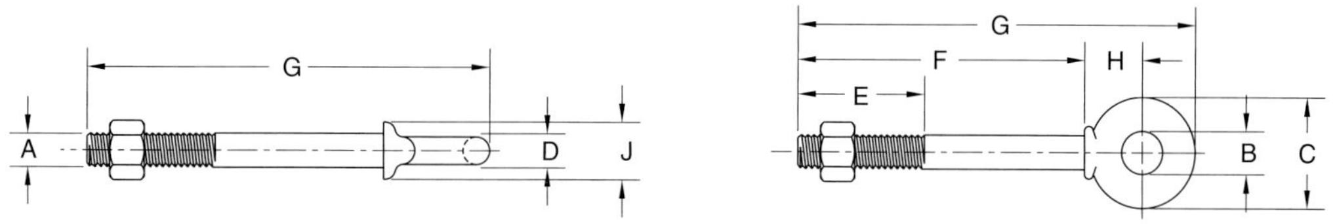

Shank Diameter and Length*	G-277 Stock No. Galv	Working Load Limit†	Weight per 100†	Dimensions*								
				A	B	C	D	E	F	G	H	J
¼ × 2	1045014	650	6.61	0.25	0.50	0.88	0.19	1.50	2.00	2.94	0.50	0.47
¼ × 4	1045032	650	8.61	0.25	0.50	0.88	0.19	2.50	4.00	4.94	0.50	0.47
⁵⁄₁₆ × 2¼	1045050	1200	12.50	0.31	0.62	1.12	0.25	1.50	2.25	3.50	0.69	0.56
⁵⁄₁₆ × 4¼	1045078	1200	18.75	0.31	0.62	1.12	0.25	2.50	4.25	5.50	0.69	0.56
⅜ × 2½	1045096	1550	19.00	0.38	0.75	1.38	0.31	1.50	2.50	3.97	0.78	0.66
⅜ × 4½	1045112	1550	31.58	0.38	0.75	1.38	0.31	2.50	4.50	5.97	0.78	0.66
½ × 3¼	1045130	2600	37.50	0.50	1.00	1.75	0.38	1.50	3.25	5.12	1.00	0.91
½ × 6	1045158	2600	56.25	0.50	1.00	1.75	0.38	3.00	6.00	7.88	1.00	0.91
⅝ × 4	1045176	5200	75.00	0.62	1.25	2.25	0.50	2.00	4.00	6.44	1.31	1.12
⅝ × 6	1045194	5200	100.25	0.62	1.25	2.25	0.50	3.00	6.00	8.44	1.31	1.12
¾ × 4½	1045210	7200	125.00	0.75	1.50	2.75	0.62	2.00	4.50	7.44	1.56	1.38
¾ × 6	1045238	7200	150.00	0.75	1.50	2.75	0.62	3.00	6.00	8.94	1.56	1.38
⅞ × 5	1045256	10,650	225.00	0.88	1.75	3.25	0.75	2.50	5.00	8.46	1.84	1.56
1 × 6	1045292	10,650	375.00	1.00	2.00	3.75	0.88	3.00	6.00	9.97	2.09	1.81
1 × 9	1045318	13,300	429.00	1.00	2.00	3.75	0.88	4.00	9.00	12.97	2.09	1.81
1¼ × 8	1045336	13,300	650.00	1.25	2.50	4.50	1.00	4.00	8.00	12.72	2.47	2.28
1¼ × 12	1045354	21,000	775.00	1.25	2.50	4.50	1.00	4.00	12.00	16.72	2.47	2.28
1½ × 15	1045372	24,000	1425.00	1.50	3.00	5.50	1.25	6.00	15.00	20.75	3.00	2.75

* in in.
† in lb

RATED LOADS* OF SHOULDERED EYEBOLTS

Eyebolt Size†	Angle of Pull			
	0°	30°	45°	60°
¼	650	420	195	160
⁵⁄₁₆	1200	780	360	300
⅜	1550	1000	465	380
½	2600	1690	780	650
⅝	5200	3380	1560	1300
¾	7200	4680	2160	1800
⅞	10,600	6890	3180	2650
1	13,300	8645	3990	3325
1¼	21,000	13,650	6300	5250
1½	24,000	15,600	7200	6000

* in lb
† in in.

RATED LOADS* OF MASTER LINKS

Nominal Size†	Material			
	End	Ring	Pear Shape	
	Alloy Steel	Carbon Steel	Alloy Steel	Carbon Steel
½	7000	—	7000	2900
⅝	9000	—	9000	4200
¾	12,300	—	12,300	6000
⅞	15,000	7200	15,000	8300
1	24,360	10,800	24,360	10,800
1⅛	—	10,400	30,600	—
1¼	36,000	17,000	36,000	16,750
1⅜	—	19,000	43,000	20,500
1½	54,300	—	54,300	—
1⅝	—	—	62,600	—
1¾	84,900	—	84,900	—
2	102,600	—	102,600	—

* in lb, with a safety factor of 5
† in in.

MACHINERY EYEBOLTS

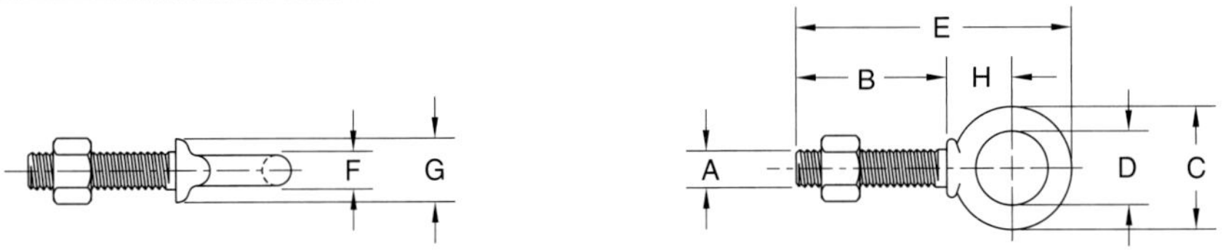

Shank Diameter and Length*	Stock No.	Working Load Limit†	Weight per 100†	Dimensions*							
				A	B	C	D	E	F	G	H
¼ × 1	9900182	650	3.20	0.25	1.00	0.88	0.50	1.94	0.19	0.47	0.50
⁵⁄₁₆ × 1⅛	9900191	1200	6.20	0.31	1.13	1.12	0.62	2.38	0.25	0.56	0.69
⅜ × 1¼	9900208	1550	12.50	0.38	1.25	1.38	0.75	2.72	0.31	0.66	0.78
½ × 1¼	9900217	2600	25.00	0.50	1.50	1.75	1.00	3.38	0.38	0.91	1.00
⅝ × 1¾	9900226	5200	50.00	0.63	1.75	2.25	1.25	4.19	0.50	1.12	1.31
¾ × 2	9900235	7200	87.50	0.75	2.00	2.75	1.50	4.94	0.62	1.38	1.56
⅞ × 2¼	9900244	10,600	150.00	0.88	2.25	3.25	1.75	5.72	0.75	1.56	1.84
1 × 2½	9900253	13,300	218.00	1.00	2.50	3.75	2.00	6.47	0.88	1.81	2.09
1¼ × 3	9900262	21,000	380.00	1.25	3.00	4.50	2.50	7.72	1.00	2.28	2.47
1½ × 3½	9900271	24,000	700.00	1.50	3.50	5.00	3.00	9.25	1.25	2.75	3.00

* in in.
† in lb

RATED LOADS* OF HOIST HOOKS WITH EYE ATTACHMENT

Throat Opening†	Material	
	Carbon Steel	Alloy Steel
0.88	1500	2000
0.97	2000	3000
1.00	3000	4000
1.12	4000	6000
1.36	6000	10,000
1.50	10,000	14,000
1.75	15,000	22,000
1.91	20,000	30,000
2.75	30,000	44,000

* in lb, with a safety factor of 5
† in in.

RATED LOADS* OF ANCHOR SHACKLES

Nominal Size†	Inside Width at Pin†	Pin Diameter†	Material	
			Carbon Steel	Alloy Steel
3/16	0.38	0.25	667	—
1/4	0.47	0.31	1000	—
5/16	0.53	0.38	1500	—
3/8	0.66	0.44	2000	4000
7/16	0.75	0.50	3000	5200
1/2	0.81	0.63	4000	6600
5/8	1.06	0.75	6500	10,000
3/4	1.25	0.88	9500	14,000
7/8	1.44	1.00	13,000	19,000
1	1.69	1.13	17,000	25,000
1 1/8	1.81	1.25	19,000	30,000
1 1/4	2.03	1.38	24,000	36,000
1 3/8	2.25	1.50	27,000	42,000
1 1/2	2.38	1.63	34,000	60,000

* in lb, with a safety factor of 5
† in in.

CHAIN SPECIFICATIONS

Nominal Chain Size*	Rated Load†		Material Diameter*	Length*	Width*	Approximate Weight† per 100'
	Grade 80	Grade 100				
7/32	2100	2700	0.217	0.67	0.32	54
9/32	3500	4300	0.275	0.88	0.41	73
5/16	4500	5700	0.330	1.02	0.48	100
3/8	7100	8800	0.397	1.22	0.57	148
1/2	12,000	15,000	0.520	1.56	0.75	250
5/8	18,100	22,600	0.630	1.93	0.92	377
3/4	28,300	35,300	0.787	2.23	1.07	570
7/8	34,200	42,700	0.875	2.25	1.14	730
1	47,700	—	1.000	3.07	1.49	985
1 1/4	72,300	—	1.260	3.92	1.74	1570

* in in.
† in lb

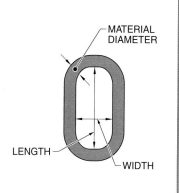

MATERIAL DIAMETER

LENGTH

WIDTH

RATED LOADS* OF CHAIN SLINGS

Grade	Nominal Chain Size†	Single-Leg Vertical	Two-Leg Sling			Four-Leg Sling		
			60°	45°	30°	60°	45°	30°
80	7/32	2100	3600	3000	2100	5500	4400	3200
	9/32	3500	6100	4900	3500	9100	7400	5200
	5/16	4500	7800	6400	4500	11,700	9500	6800
	3/8	7100	12,300	10,000	7100	18,400	15,100	10,600
	1/2	12,000	20,800	17,000	12,000	31,200	25,500	18,000
	5/8	18,100	31,300	25,600	18,100	47,000	38,400	27,100
	3/4	28,300	49,000	40,000	28,300	73,500	60,000	42,400
	7/8	34,200	59,200	48,400	34.200	88,900	72,500	51,300
	1	47,700	82,600	67,400	47,700	123,900	101,200	71,500
	1 1/4	72,300	125,000	102,200	72,300	187,800	153,400	108,400
100	7/32	2700	4700	3800	2700	7000	5700	4000
	9/32	4300	7400	6100	4300	11,200	9100	6400
	5/16	5700	9900	8100	5700	14,800	12,600	8500
	3/8	8800	15,220	12,400	8800	22,900	18,700	13,200
	1/2	15,000	26,000	21,200	15,000	39,000	31,800	22,500
	5/8	22,600	39,100	32,000	22,600	58,700	47,900	33,900
	3/4	35,300	61,100	49,900	35,300	91,700	74,900	53,000
	7/8	42,700	74,000	60,400	42,700	110,900	90,600	64,000

* in lb
† in in.

RATED LOADS* OF WEB SLINGS

Class	Plies	Width†	Vertical	Choker	Bridle or Basket Vertical	Bridle or Basket 60°	Bridle or Basket 45°	Bridle or Basket 30°	Type V Endless
5	1	1	1100	880	2200	1900	1600	1100	2200
		1½	1600	1280	3200	2800	2300	1600	3200
		1¾	1900	1520	3800	3300	2700	1900	3800
		2	2200	1760	4400	3800	3100	2200	4400
		3	3300	2640	6600	5700	4700	3300	6600
		4	4400	3520	8800	7600	6200	4400	8800
		5	5500	4400	11,000	9500	7800	5500	11,000
		6	6600	5280	13,200	11,400	9300	6600	13,200
	2	1	2200	1760	4400	3800	3100	2200	4400
		1½	3300	2640	6600	5700	4700	3300	6600
		1¾	3800	3040	7600	6600	5400	3800	7600
		2	4400	3520	8800	7600	6200	4400	8800
		3	6600	5280	13,200	11,400	9300	6600	13,200
		4	8200	6560	16,400	14,200	11,600	8200	16,400
		5	10,200	8160	20,400	17,700	14,400	10,200	20,400
		6	12,300	9840	24,600	21,300	17,400	12,300	24,600
7	1	1	1600	1280	3200	2800	2300	1600	3200
		1½	2300	1840	4600	4000	3300	2300	4600
		1¾	2700	2160	5400	4700	3800	2700	5400
		2	3100	2480	6200	5400	4400	3100	6200
		3	4700	3760	9400	8100	6600	4700	9400
		4	6200	4960	12,400	10,700	8800	6200	12,400
		5	7800	6240	15,600	13,500	11,000	7800	15,600
		6	9300	7440	18,600	16,100	13,200	9300	18,600
		8	11,800	9440	23,600	20,400	16,700	11,800	23,600
		10	14,700	11,760	29,400	25,500	20,800	14,700	29,400
		12	17,600	14,080	35,200	30,500	24,900	17,600	35,200
	2	1	3100	2480	6200	5400	4400	3100	6200
		1½	4700	3760	9400	8100	6600	4700	9400
		1¾	5400	4320	10,800	9400	7600	5400	10,800
		2	6200	4960	12,400	10,700	8800	6200	12,400
		3	8800	7040	17,600	15,200	12,400	8800	17,600
		4	11,000	8800	22,000	19,100	15,600	11,000	22,000
		5	13,700	10,960	27,400	23,700	19,400	13,700	27,400
		6	16,500	13,200	33,000	28,600	23,000	16,500	33,000
		8	22,700	18,160	45,400	39,300	32,100	22,700	45,400
		10	28,400	22,720	56,800	49,200	40,200	28,400	56,800
		12	34,100	27,280	68,200	59,100	48,200	34,100	68,200
	4	1	5500	4400	11,000	9500	7800	5500	
		2	11,000	8800	22,000	19,100	15,600	11,000	
		3	16,400	13,120	32,800	28,400	23,200	16,400	
		4	20,400	16,320	40,800	35,300	28,800	20,400	—
		5	25,500	20,400	51,000	44,200	36,100	25,500	
		6	30,600	24,480	61,200	53,000	43,300	30,600	

* in lb
† in in.

RATED LOADS* OF ROUNDSLINGS

		Vertical	Choker	Basket			
				Vertical	60°	45°	30°
Size	Color Code						
1	Purple	2600	2100	5200	4500	3700	2600
2	Green	5300	4200	10,600	9200	7500	5300
3	Yellow	8400	6700	16,800	14,500	11,900	8400
4	Tan	10,600	8500	21,200	18,400	15,000	10,600
5	Red	13,200	10,600	26,400	22,900	18,700	13,200
6	White	16,800	13,400	33,600	29,100	23,800	16,800
7	Blue	21,200	17,000	42,400	36,700	30,000	21,200
8	Orange	25,000	20,000	50,000	43,300	35,400	25,000
9	Orange	31,000	24,800	62,000	53,700	43,800	31,000
10	Orange	40,000	32,000	80,000	69,300	56,600	40,000
11	Orange	53,000	42,400	106,000	91,800	74,900	53,000
12	Orange	66,000	52,800	132,000	114,300	93,000	66,000
13	Orange	90,000	72,000	180,000	155,900	127,300	90,000

*in lb

BREAKING STRENGTHS* OF SELECTED WIRE ROPES

Diameter[†]	Improved Plow Steel		Extra-Improved Plow Steel
	Fiber Core	IWRC[‡]	IWRC[‡]
1/4	5340	5740	6640
5/16	8300	8940	10,280
3/8	11,900	12,800	14,720
7/16	16,120	17,340	19,900
1/2	20,800	22,400	26,000
9/16	26,400	28,200	32,800
5/8	32,600	35,000	40,200
3/4	46,400	50,000	57,400
7/8	62,800	67,400	77,600
1	81,600	87,600	100,800

* in lb, for uncoated, general purpose, rotation-resistant 6 × 19 (class 2) or
  6 × 37 (class 3) wire rope

† in in.

‡ independent wire rope core

BREAKING STRENGTHS* OF SELECTED FIBER ROPES

Diameter†	Manila 3-Strand Twisted	Polypropylene 3-Strand Twisted	Polyester			Nylon		
			3-Strand Twisted	Double Braided	8-Strand Plaited	3-Strand Twisted	Double Braided	8-Strand Plaited
⅝	405	720	800	—	—	950	—	—
¾	540	1130	1200	1900	2000	1500	1785	1500
1	900	1710	2000	2935	3100	2600	2835	2500
1⅛	1215	2440	2800	4245	4500	3300	4095	3700
1¼	1575	3160	3800	5730	6000	4800	5355	5000
1½	2385	3780	5000	7500	7700	5800	7245	6400
1¾	3105	4600	6500	9450	9700	7600	9450	8000
2	3960	5600	8000	11,660	12,100	9800	12,600	11,000

* in lb
† in in.

ELECTRIC HOIST INSPECTION CHECKLISTS

Item	Daily	Monthly	Semi-annually	Deficiencies
All functional operating mechanisms	✓	✓	✓	Maladjustment interfering with proper operation, excessive component wear
Controls	✓		✓	Improper operation
Safety devices	✓		✓	Malfunction
Hooks	✓	✓	✓	Deformation, chemical damage, 15% in excess of normal throat opening, 10% twist from plane of unbent hook, cracks
Load-bearing components (except rope or chain)	✓	✓	✓	Damage (especially if hook is twisted or pulling open)
Load-bearing rope	✓	✓	✓	Wear, twist, distortion, improper dead-ending, deposits of foreign material, broken wires
Load-bearing chain	✓	✓	✓	Wear, twist, distortion, improper dead-ending, deposits of foreign material
Fasteners	✓	✓	✓	Not tight
Drums, pulleys, sprockets			✓	Cracks, excessive wear
Pins, bearings, shafts, gears, rollers, locking and clamping devices			✓	Cracks, excessive wear, distortion, corrosion
Brakes	✓		✓	Excessive wear, drift
Electrical			✓	Pitting, loose wires
Contactors, limit switches, pushbutton stations			✓	Deterioration, contact wear, loose wires
Hook retaining members (collars, nuts) and pins, welds, or rivets securing them			✓	Not tight or secure
Supporting structure or trolley			✓	Continued ability to support imposed loads
Warning label	✓		✓	Removed or illegible
Pushbutton markings	✓		✓	Removed or illegible
Capacity marking	✓		✓	Removed or illegible

RIGGING FORMULAS

CONDUCTOR WEIGHT ON POLE

$$\text{conductor weight} = \frac{\text{span a} + \text{span b}}{2} \times \text{weight/ft} \times \text{safety factor}$$

CONDUCTOR WEIGHT ON POLE

$$\text{line tension} = \frac{\text{span}^2 \times \text{weight/ft}}{(8 \times \text{sag})}$$

CONDUCTOR BISECT TENSION

$$\text{bisect tension} = \frac{\text{bisect line length}}{\text{distance}} \times \text{line tension} \times 2$$

COMPRESSIVE FORCE ON GUYED POLE

$$\text{compressive force} = \frac{\text{line tension} \times \text{pole height}}{\text{anchor distance}}$$

GUY TENSION

$$\text{guy tension} = \frac{\text{line tension} \times \text{guy length}}{\text{anchor distance}}$$

GUY LENGTH

$$\text{guy length} = \sqrt{\text{anchor distance}^2 + \text{pole height}^2}$$

PYTHAGOREAN THEOREM

$$A^2 + B^2 = C^2$$

SAG

$$\text{sag} = \frac{\text{weight/ft} \times \text{span}^2}{8 \times \text{tension}}$$

PULL REQUIRED ON FALL LINE

$$\text{Pull} = \frac{\text{weight} \times \text{number of parts of rope}}{\text{mechanical advantage}}$$

ALUMINUM CONDUCTOR STEEL REINFORCED TABLE

ACSR

Code Word	Size (AWG or kcmil)	Strand-ing (Al/Stl)	Diameter (ins.) Individual Wires Al	Diameter (ins.) Individual Wires Stl	Steel Core	Complete Cable	Weight Per 1000 ft. (lbs.) Al	Weight Per 1000 ft. (lbs.) Stl	Weight Per 1000 ft. (lbs.) Total	Content (%) Al	Content (%) Stl	Rated Strength (lbs.)	Resistance OHMS/1000 ft. DC @ 20°C	Resistance OHMS/1000 ft. AC @ 75°C	Allowable Ampacity (Amps)
Turkey	6	6/1	.0661	.0661	.0661	.198	24	12	36	67.88	32.12	1190	.641	.806	105
Swan	4	6/1	.0834	.0834	.0834	.25	39	18	57	67.87	32.12	1860	.403	.515	140
Swanate	4	7/1	.0772	.103	.103	.257	39	28	67	58.1	41.9	2360	.399	.519	140
Sparrow	2	6/1	.1052	.1052	.1052	.316	62	29	91	67.9	32.1	2850	.254	.332	184
Sparate	2	7/1	.0974	.1298	.1298	.325	62	45	107	58.12	41.88	3460	.251	.338	184
Robin	1	6/1	.1181	.1181	.1181	.354	78	37	115	67.88	32.12	3550	.201	.268	212
Raven	1/0	6/1	.1327	.1327	.1327	.398	99	47	145	67.89	32.11	4380	.159	.217	242
Quail	2/0	6/1	.1489	.1489	.1489	.447	124	59	183	67.88	32.12	5310	.126	.176	276
Pigeon	3/0	6/1	.1672	.1672	.1672	.502	156	74	230	67.87	32.13	6620	.100	.144	315
Penguin	4/0	6/1	.1878	.1878	.1878	.563	197	93	291	67.88	32.12	8350	.0795	.119	357
Waxwing	266.8	18/1	.1217	.1217	.1217	.609	250	39	289	86.43	13.57	6880	.0643	.0787	449
Partridge	266.8	26/7	.1013	.0788	.2363	.642	251	115	367	68.51	31.49	11300	.0637	.0779	475
Ostrich	300	26/7	.1074	.0835	.2506	.68	283	130	412	68.51	31.49	12700	.0567	.0693	492
Merlin	336.4	18/1	.1367	.1367	.1367	.684	315	49	365	86.43	13.57	8680	.0510	.0625	519
Linnet	336.4	26/7	.1137	.0885	.2654	.72	317	146	462	68.51	31.49	14100	.0505	.0618	529
Oriole	336.4	30/7	.1059	.1059	.3177	.741	318	209	526	60.35	39.65	17300	.0502	.0613	535
Chickadee	397.5	18/1	.1486	.1486	.1486	.743	373	58	431	86.43	13.57	9940	.0432	.0529	576
Brant	397.5	24/7	.1287	.0858	.2574	.772	374	137	511	73.21	26.79	14600	.0430	.0526	584
Ibis	397.5	26/7	.1236	.0962	.2885	.783	374	172	546	68.51	31.49	16300	.0428	.0523	587
Lark	397.5	30/7	.1151	.1151	.3453	.806	375	247	622	60.35	39.65	20300	.0425	.0519	594
Pelican	477	18/1	.1628	.1628	.1628	.814	447	70	517	86.44	13.56	11800	.0360	.0442	646
Flicker	477	24/7	.141	.094	.2819	.846	449	164	614	73.21	26.79	17200	.0358	.0439	655
Hawk	477	26/7	.1354	.1053	.316	.858	449	207	656	68.51	31.49	19500	.0356	.0436	659
Hen	477	30/7	.1261	.1261	.3783	.883	450	296	746	60.35	39.65	23800	.0354	.0433	666
Osprey	556.5	18/1	.1758	.1758	.1758	.879	522	82	603	86.43	13.57	13700	.0308	.0379	711
Parakeet	556.5	24/7	.1523	.1015	.3045	.914	524	192	716	73.21	26.79	19800	.0307	.0376	721
Dove	556.5	26/7	.1463	.1138	.3413	.927	524	241	765	68.51	31.49	22600	.0306	.0375	726
Eagle	556.5	30/7	.1362	.1362	.4086	.953	525	345	871	60.35	39.65	27800	.0303	.0372	734
Peacock	605	24/7	.1588	.1059	.3177	.953	570	209	779	73.2	26.8	21600	.0282	.0346	760
Squab	605	26/7	.1525	.1186	.3559	.966	570	262	832	68.51	31.49	24300	.0281	.0345	765
Wood Duck	605.0	30/7	.142	.142	.426	.994	571	375	946	60.35	39.65	28900	.0279	.0342	774
Teal	605.0	30/19	.142	.0852	.426	.994	571	367	939	60.86	39.14	30000	.0279	.0342	773
Kingbird	636	18/1	.188	.188	.188	.94	596	94	690	86.43	13.57	15700	.0270	.0332	773
Swift	636.0	36/1	.1329	.1329	.1329	.93	596	47	643	92.72	7.28	13690	.0271	.0334	769
Rook	636	24/7	.1628	.1085	.3256	.977	599	219	818	73.22	26.78	22600	.0268	.0330	784
Grosbeak	636	26/7	.1564	.1216	.3649	.991	599	275	874	68.51	31.49	25200	.0267	.0328	789

Southwire
One Southwire Drive
Carrollton, Ga. 30119 USA

MADE
IN ★ THE
USA

ARCHITECTURAL SYMBOLS . . .

Material	Elevation	Plan View	Section
Earth			
Brick	With note indicating Type of brick (Common, face, etc.)	Common or face / Firebrick	Same as plan views
Concrete		Lightweight / Structural	Same as plan views
Concrete Masonry Units		Or	Or
Stone	Cut stone / Rubble	Cut stone / Rubble / Cast stone (concrete)	Cut stone / Cast stone (Concrete) / Rubble or Cut stone
Wood	Siding / Panel	Wood stud / Remodeling / Display	Rough Members / Finished Members / Plywood
Plaster		Wood stud, lath, and plaster / Metal lath and plaster / Solid plaster	Lath and Plaster
Roofing	Shingles	Same as elevation	
Glass	Or / Glass block	Glass / Glass block	Small Scale / Large Scale

. . . ARCHITECTURAL SYMBOLS

Material	Elevation	Plan View	Section
Facing Tile	Ceramic Tile	Floor Tile	Ceramic Tile Large Scale Ceramic Tile Small Scale
Structural Clay Tile			Same as Plan View
Insulation		Loose Fill or Batts Rigid Spray Foam	Same as Plan View
Sheet Metal Flashing		Occasionally Indicated by Note	
Metals Other than Flashing	Indicated by Note or Drawn to Scale	Same as Plan View	Small Scale Steel Cast Iron Aluminum Bronze or Brass
Structural Steel	Indicated by Note or Drawn to Scale	or	Rebars Small Scale Large Scale L-ANGLES, S-BEAMS, ETC.

PLOT PLAN SYMBOLS

North	Fire Hydrant	Walk	E / OR	Electric Service
Point of Beginning (POB)	Mailbox	Improved Road	G / OR	Natural Gas Line
Utility Meter or Valve	Manhole	Unimproved Road	W / OR	Water Line
Power Pole and Guy	Tree	Building Line	T / OR	Telephone Line
Light Standard	Bush	Property Line		Natural Grade
Traffic Signal	Hedgerow	Property Line		Finish Grade
Street Sign	Fence	Township Line	+ XX.00′	Existing Elevation

Glossary

A

adjustable transformer sling: A sling with adjustable eyes at each end to accommodate various types and sizes of transformers.

air caster: A pneumatic device that is placed under a load to allow horizontal movement of the load.

alloy: A metal formulated from the combination of two or more elements.

angle of loading: *See* sling angle.

anti-two-block device: A safety device mounted close to the fixed end of a crane's hoisting line that either sounds an alarm or immediately stops the crane and hoist if it is touched by the hoist hook assembly.

articulating boom crane: A mobile crane that has a boom with a hinged joint to allow the boom to pivot and is typically mounted on commercial trucks. Also known as a knuckle boom crane.

audible signal: A signal given by a signalperson with a sound-making device indicating a desired crane movement to a crane operator.

B

basket hitch: A sling hitch in which a sling is passed under a load, and both ends of the sling are connected to a hoist hook.

beam clamp: A fixed device for attaching a hoist to the bottom of a beam at a single location.

becket: An attachment point, usually on a block, for the dead end of a hoisting rope.

bend: A knot that is used to tie the ends of two ropes together.

bending efficiency: The ratio of the strength of a bent rope to its nominal strength rating.

bend ratio: The ratio of the diameter of a bend to the nominal diameter of the rope.

bight: A loose or slack part of a rope between two ends.

bird caging: A type of damage to wire rope in which the outer strands separate and open.

block: An assembly of one or more sheaves in a frame.

block and tackle: A combination of sheaves and ropes used to improve lifting efficiency.

block loading: The total amount of static force experienced by a block while in a certain arrangement.

boom: A long beam that projects from the main part of a crane in order to extend the reach of the hoist.

boom angle: The angle between a horizontal plane and a boom.

boom deflection: The bending down of a boom due to the weight of a load.

bowline on a bight: A bowline knot tied in the middle of a line to tie the line around an object.

braiding: The weaving of three or more untwisted strands into a rope.

breaking strength: The tension at which a material is expected to break.

bridge crane: *See* overhead crane.

bridle hitch: A sling hitch in which two or more slings share a common end fitting that is used as a lifting point.

bridle sling: An assembly of two or more sling legs, each with an end gathered together at a common end fitting.

bull rope: Heavy rope that is used to pull wires.

C

cab: A compartment or platform attached to a crane on which an operator rides.

cabling: The tendency of a rope to rotate and untwist when under load.

cantilever: A projecting structure supported at only one end.

capsizing: The tendency of a knot under load to change form, often in an adverse way.

capstan hoist: A portable winchlike device with a rotating drum (capstan) used for lifting.

carabiner: An oblong metal ring with one spring-hinged side that is used in climbing as a connector and to hold a freely running rope.

center of gravity (CG): The point in space at which an object's mass is considered to be concentrated. Also known as the center of mass.

center of mass: *See* center of gravity.

chain: A series of connected metal links.

choke angle: The angle formed at the choke between the vertical part of a sling and the part of the sling surrounding a load.

choker hitch: A sling hitch in which one end of a sling is wrapped under or around a load, passed through the eye at the other end of the sling, and then connected to a hoist hook.

choker hook: A sliding hook used to form a choker sling when hooked to a sling eye.

climbing tower crane: A tower crane that is secured to the frame of a high-rise structure being erected and can be periodically raised as new floor levels are added to the structure.

code: A collection of regulations related to a particular trade or environment.

competent person: According to OSHA, "one who is capable of identifying existing and predictable hazards in the surroundings or working conditions which are unsanitary, hazardous, or dangerous to employees, and who has authorization to take prompt corrective measures to eliminate them."

compression: The inward pushing force on an object.

core: A strand of metal wire or fiber that forms the center of a wire rope.

corrosion: The disintegration of a material due to a chemical reaction with its environment.

cosecant: A trigonometric function equal to the ratio between the hypotenuse and the opposite legs of a right triangle.

counterweight: A block of dense material attached to one end of a structure in order to offset the moment force of the weight attached to the other end of the structure.

crane: A combination of a hoist and a structure used to support and move a load.

crane block: A block used in the hook assembly of a crane. Also known as a hook block.

cribbing: Blocking used to temporarily support a load while at rest.

D

dead end: The loose end of a rope.

design factor: *See* safety factor.

digger derrick: A mobile telescoping-boom crane with an auger attachment that is typically mounted on a utility truck.

dolly: A wheeled platform, sometimes with a handle, that is placed under a load to allow horizontal movement.

double bowline: A bowline knot that has an extra loop.

dressing: The adjustment of a knot as it is tightened for a clean look and tight arrangement.

drift: The tendency of a load to continue moving after the crane has stopped.

dynamometer: A device that measures linear force by measuring the rotational force (torque) that it induces on a load cell sensor.

E

endless sling: A sling formed by attaching the ends of a sling body together to form a continuous loop. Also called a grommet sling.

end truck: A roller assembly consisting of a frame, wheels, and bearings.

energy: The capacity to do work.

equalizer beam: A load leveler that is used to stabilize a load by distributing the load equally between two sling legs or equalizing loads on two hoist lines when performing a tandem lift.

eye-and-eye sling: A sling in which loops are formed at each end of the sling body by doubling over the material and securing it by sewing, weaving, or using a compression fitting.

eyebolt: A bolt with a looped head that is fastened to a load to provide a lift point.

eyenut: A loop-shaped nut that is fastened to a load to provide a lift point.

F

fabrication efficiency: The ratio of the tensile strength of a webbing material to the tensile strength of the web sling into which it is fabricated.

fall block: *See* traveling block.

fall zone: An area where partially or completely suspended materials might land if they become loose from a load or if the rigging or crane fails.

freestanding tower crane: A tower crane that is secured to a concrete foundation next to the structure being erected.

friction: The force of two objects or surfaces resisting movement.

force: The interaction between two objects.

forklift: A vehicle with hydraulically operated forks used to lift and transport loads.

foundry hook: A hook with a wide, deep throat that fits the handles of molds or casting.

fracture: A crack in metal caused by the stress and fatigue of repeated pulling or bending forces.

frequent inspection: A rigging equipment inspection performed at the beginning of each workday or shift by the user of the equipment or another designated person.

friction: The force of two objects or surfaces resisting movement.

G

gantry crane: An industrial crane composed of a hoist trolley that travels along a horizontal bridge beam supported by a leg assembly.

gate block: *See* snatch block.

gin block: A simple block that is not much more than a large sheave in a lightweight frame. Also known as a well wheel.

grab hook: A hook that can engage and securely hold a chain link.

grip: A mechanical device used for pulling cable, wire, or rope.

H

hand line: A rope, block, and set of clips or hooks that can be attached to the crossarm of a utility pole.

hand signal: A visual signal given by a signalperson using their hands indicating a desired crane movement to a crane operator.

helicopter crane: A helicopter that is specially designed to lift heavy loads with a long hoist or sling.

hitch: The binding of rope to another object, often temporarily.

hoist hook: A hook with a rounded shape that is suitable for most rigging and lifting applications.

hoisting: The transportation of a load by a crane or hoist.

hook: A curved implement used for quickly and temporarily connecting rigging to loads or lifting equipment.

hook block: *See* crane block.

horizontal angle: *See* sling angle.

I

industrial crane: An indoor crane with permanent structural beam supports.

initial inspection: A rigging equipment inspection performed before equipment is placed into service in which a designated person ensures that the equipment meets the applicable government regulations and industry standards.

J

jib crane: An industrial crane composed of a hoist trolley that travels along a horizontal boom, which is supported by a single structural leg.

Johnson bar: *See* lever dolly.

K

kinking: A sharp bend that permanently deforms the lay of rope strands.

knot: The interlacing of a part of a rope to itself, which is then drawn tight.

knot efficiency: The ratio of the strength of a knotted rope to its nominal strength rating.

knuckle boom crane: *See* articulating boom crane.

L

lacing: A type of reeving in which a rope starts on one side of blocks and passes to the other side one sheave at a time.

lang-lay rope: Rope in which the wires and strands are twisted in the same direction.

lashing: A method of binding two objects to each other by wrapping rope around them numerous times.

latch: A load-operated jaw, rotating gate, or spring-loaded clip that closes off the opening of a hook.

lattice-boom crane: A mobile crane with a boom constructed from one or more sections of thin steel gridwork.

lay: 1. The length of rope in which a strand makes one complete spiral wrap or braiding/plaiting pattern. **2.** A designation for the direction in which rope strands are twisted, described as if spiraling away from the observer.

lead line: The part of a rope to which force is applied to hold or move a load.

left-hand lay rope: Rope with strands that spiral to the left (counterclockwise).

lever: A bar that turns on a point or fulcrum.

lever dolly: A pry bar with wheels that is used to raise a load slightly or reposition a barely hoisted load. Also called a Johnson bar.

lift: One complete set of hoisting actions performed by workers who transport a load from a starting point to a destination.

lifting beam: A bar, beam, or tube used to change the direction of rigging forces at the load from angled to vertical. Also known as a spreader beam.

lift plan: An evaluation of the potential hazards of a lift and the equipment and procedures required to safely execute the lift.

link: A plain, rigid loop attached to the end of a sling or hung on a crane hook that helps connect multiple rigging components together.

live end: The load-lifting end of a rope.

load: An object that must be transported via hoisting.

load capacities chart: The manufacturer-rated maximum weight that a digger derrick can lift under certain circumstances.

load cell: A device that measures linear force with an electrical transducer.

load path: The path a load must take from its starting point to its destination and that includes every point in between.

load radius: The horizontal distance from the pivot point of a crane to a point below the hoist hook.

load triangle: The right triangle formed by an angled sling and the horizontal and vertical forces acting on the sling.

loop: The folding or doubling of a line to create an opening through which another line may pass.

loop eye: A length of webbing folded back and spliced to the sling body, forming a closed loop.

M

machinery lift: A two-wheeled dolly with swivel casters and a hydraulic lift, which is placed at the edge of a load.

material-handling bucket truck: An aerial lift with a material-handling jib that is capable of lifting light loads.

mechanical advantage: The ratio of the output force from a machine to the input force applied to the machine.

mobile crane: A crane that can be moved within and between job sites.

moment: The tendency of a force to rotate an object around a point.

N

nip: A pressure point created when a rope crosses over itself after a turn around an object.

nominal value: A specified value that may vary slightly from the actual value.

O

one-handed bowline: A bowline knot that is tied with only one hand.

outrigger: A structure that extends out from the platform of a mobile crane and contacts the ground with a large pad.

overhead crane: An industrial crane composed of a hoist trolley that travels along a horizontal bridge beam, which travels along a pair of overhead runways. Also known as a bridge crane.

P

pallet: An open structure, typically made of wood, that supports a load and provides openings for lifting equipment to be easily placed under the load.

pallet jack: A dolly consisting of two forks that are hydraulically lifted above a set of wheels.

part: A rope length between a dead end and block or between two blocks.

pendant: A control unit that hangs down from the hoists of smaller industrial cranes.

periodic inspection: A thorough rigging equipment inspection performed at regular intervals.

plaiting: The weaving of four pairs of alternately twisted strands into a rope.

power: The rate of doing work or using energy.

preformed rope: Wire rope in which the strands are permanently formed into a helical shape during fabrication.

proof test: A nondestructive test in which a sling is subjected to a tension force greater than its rated load but less than its breaking strength.

pry bar: A forged steel bar with an angled and flattened end that is pushed under the edge of a load in order to lift the load slightly. Also known as a pinch bar.

pulley: A simple machine that consists of a grooved wheel attached to a frame or block.

pulling rope: A high-strength, low-stretch rope that is used for tension stringing conductors.

Q

qualified person: According to OSHA, "one who, by possession of a recognized degree, certificate, or professional standing, or who by extensive knowledge, training and experience, has successfully demonstrated the ability to solve or resolve problems relating to the subject matter, the work, or the project."

R

range diagram: A manufacturer chart used to determine digger derrick load radius based on the boom angle and extension.

rated capacity: The maximum tension that a crane or hoist may be subjected to while maintaining the appropriate margin of safety.

rated load: The maximum tension that a rigging component may be subjected to while maintaining an appropriate margin of safety.

reach: The distance from the inside of a sling's upper end fitting to the inside of its lower end fitting.

reeving: The threading of a rope through an opening or around a sheave.

regular-lay rope: Rope in which the wires in the strands are twisted in the opposite direction of the lay of the strands.

regulation: A rule made mandatory by a federal, state, or local government.

return eye sling: A sling that has multiple widths of webbing held edge to edge and in which each eye forms a right angle to the plane of the sling body. Also known as a reverse eye sling.

reverse eye sling: *See* return eye sling.

rigging: The assembly of components that connect a load to the lifting hook of a crane or hoist.

rigging dolly: A heavy-duty dolly, often with rollers, used to move large loads.

rigging hardware: The category of components used to connect loads to slings, slings to slings, and slings to hoist hooks.

right-hand lay rope: Rope with strands that spiral to the right (clockwise).

rope: A length of fibers or thin wires that are twisted or braided together to form a strong and flexible line.

rope block: An assembly of one or more sheaves in a frame.

roundsling: An endless (continuous loop) sling made from a bundle of unwoven synthetic fiber yarns enclosed in a protective cover of synthetic fiber.

running bowline: A bowline-knot-and-noose combination that does not bind, slides easily, and can be undone easily.

runway: A rail-and-beam combination that allows for the movement of end trucks.

S

safety factor: The ratio of a component's breaking strength to its rated load. Also known as the design factor.

screw: A cylindrical shaft with spiral grooves or ridges called threads.

seizing: The wire wrapping that binds the end of a wire rope near where it is cut.

selvedge: An edge treatment on woven material that helps prevent unraveling.

shackle: A U-shaped metal connector with holes drilled into the ends for receiving a removable pin or bolt.

shear force: The pushing force on one part of an object in one direction and another part of the object in the opposite direction.

sheave: A grooved wheel attached to a frame or block that supports a rope that is changing direction. Also known as a pulley.

shock loading: The abrupt application of force to an object.

side loading: The application of a horizontal force on a boom due to the hoisting of a load that is not directly underneath the hoist hook.

signalperson: A worker who is specially trained to provide clear and standardized signals to a crane operator in order to guide the movement of a load during a lift.

skip reeving: A method of reeving in which a rope starts at a center sheave and passes from one side of blocks to the other side before being attached to a becket.

sling: A flexible length of load-bearing material that is used to rig a load.

sling angle: The acute angle between horizontal and the sling leg. Also known as the angle of loading or horizontal angle.

sling hitch: An arrangement of one or more slings for connecting a load to a hoist hook.

snatch block: A block with side plates that can be opened, allowing it to be added to a hoisting line without access to the end of the line. Also known as a gate block.

socket: A fitting attached to the end of wire rope to provide a means for making strong connections.

softener: A relatively soft material used between loads and rigging in order to limit damaging contact.

sorting hook: A hook with a straight, tapered tip that can be used to directly hold plates, cylinders, and other shapes that allow full engagement.

spelter socket: A socket that uses molten zinc or resin to secure the end of a wire rope inside the socket.

splice: 1. The unlaying and then reweaving together of two portions of rope in order to form a permanent connection. **2.** An overlap of webbing material that is sewn together.

spreader beam: *See* lifting beam.

square reeving: A method of reeving in which a rope passes between sheaves at right angles to each other, which makes the rope change direction at each sheave.

standard: A collection of voluntary rules developed through consensus and related to a particular trade, industry, or environment.

standing block: The upper block in a block-and-tackle configuration that is usually attached to a fixed object or structure.

standing end: The end of a standing part.

standing part: The portion of a rope that is unaltered or not involved in making a knot, hitch, or bend.

static force: A constant force applied to an object that is only sufficient to keep the object in place.

stopper knot: A category of knots that are used to prevent the ends of ropes from slipping through openings or other knots.

strand: A bundle of wires twisted spirally around an axis.

swaged sleeve: A compression fitting that is crimped onto the two portions of wire rope that meet at the base of an eye loop.

swage socket: A socket that is compressed onto the end of a wire rope.

swaging: A mechanical process of forming a metal into a certain shape, usually at ambient temperatures.

swivel hoist ring: A freely rotating attachment that is fastened to a load to provide a lift point.

symmetry: The characteristic of one side of an object mirroring its opposite side.

T

tackle: The combination of ropes and accessories used with blocks to gain mechanical advantage for lifting.

tackle block: A block used with natural or synthetic fiber ropes, primarily for manual hoisting operations.

tag line: A rope, handled by a qualified individual, used to control rotational movement of a load during a lift.

telescoping-boom crane: A mobile crane with an extendable boom composed of nested sections.

tension: The outward pulling force on an object.

termination efficiency: The ratio of the rated load of a wire rope sling to the rated load of the unterminated wire rope.

thimble: A curved piece of metal that supports a loop of rope and protects it from sharp bends and abrasion.

three-ring bowline: A bowline knot that is formed quickly with three loops of rope.

torque: The rotational (turning) force on an object.

tower crane: A fixed crane consisting of a high, vertical mast tower topped by a horizontal jib.

traveling block: The lower moveable block in a block-and-tackle configuration that is attached to a load, usually via a hook. Also known as a fall block.

triangle end sling: A sling with metal triangle-shaped fittings on each end.

triple bowline: A bowline knot that has three loops and is typically used to help with lifting injured personnel.

trolley: A wheeled assembly that travels horizontally along a crane beam or boom.

turnback: The portion of the end of a rope that is folded back on itself.

turnbuckle: An adjustable-length attachment used for connecting other rigging components.

twisted eye sling: An eye-and-eye sling in which each eye forms a right angle to the plane of the sling body.

two blocking: A condition in which a device on the hoisting line, such as a block or overhaul ball, is drawn up against the sheave at the end of a boom.

U

unlaying: The untwisting of a rope's strands.

V

vertical hitch: A sling hitch in which one end of a sling connects to a hoist hook and the other end connects to a load.

voice signal: A signal given vocally by a signalperson indicating a desired crane movement to a crane operator.

W

wear pad: A leather or webbed pad used to cover the body of a web sling in order to protect it from damage.

webbing: Flat, narrow strapping woven from yarns of strong synthetic fibers.

web sling: A flat rigging sling made from synthetic webbing material.

wedge socket: A socket that holds a loop of wire rope securely with a wedge that is tightened by tension on the rope.

well wheel: *See* gin block.

wheel: A circular object that rotates about its axis and is typically used in conjunction with an axle to easily move heavy objects.

whipping: The wrapping of twine that binds the end of a fiber rope near where it is cut.

wireless control box: A control unit connected wirelessly to a crane or hoist.

wire rope block: A block used with powered hoists when severe conditions of service are not expected.

work: The energy used when a force is exerted over a distance.

working end: The end of a working part.

working part: The portion of a rope involved in making a knot, hitch, or bend.

Y

yarn: A continuous line of fibers twisted together.

Index

Page numbers in italic refer to figures.